Advances in Experimental Medicine and Biology

Volume 990

Editorial Board:

IRUN R. COHEN, *The Weizmann Institute of Science*
ABEL LAJTHA, *N.S. Kline Institute for Psychiatric Research*
JOHN D. LAMBRIS, *University of Pennsylvania*
RODOLFO PAOLETTI, *University of Milan*

For further volumes:
http://www.springer.com/series/5584

Dariusz Leszczynski
Editor

Radiation Proteomics

The effects of ionizing and
non-ionizing radiation on cells
and tissues

Editor
Dariusz Leszczynski
STUK
Radiation and Nuclear Safety Authority
Helsinki, Finland

ISSN 0065-2598
ISBN 978-94-007-5895-7   ISBN 978-94-007-5896-4 (eBook)
DOI 10.1007/978-94-007-5896-4
Springer Dordrecht Heidelberg New York London

Library of Congress Control Number: 2013931102

© Springer Science+Business Media Dordrecht 2013
This work is subject to copyright. All rights are reserved by the Publisher, whether the whole or part of the material is concerned, specifically the rights of translation, reprinting, reuse of illustrations, recitation, broadcasting, reproduction on microfilms or in any other physical way, and transmission or information storage and retrieval, electronic adaptation, computer software, or by similar or dissimilar methodology now known or hereafter developed. Exempted from this legal reservation are brief excerpts in connection with reviews or scholarly analysis or material supplied specifically for the purpose of being entered and executed on a computer system, for exclusive use by the purchaser of the work. Duplication of this publication or parts thereof is permitted only under the provisions of the Copyright Law of the Publisher's location, in its current version, and permission for use must always be obtained from Springer. Permissions for use may be obtained through RightsLink at the Copyright Clearance Center. Violations are liable to prosecution under the respective Copyright Law.
The use of general descriptive names, registered names, trademarks, service marks, etc. in this publication does not imply, even in the absence of a specific statement, that such names are exempt from the relevant protective laws and regulations and therefore free for general use.
While the advice and information in this book are believed to be true and accurate at the date of publication, neither the authors nor the editors nor the publisher can accept any legal responsibility for any errors or omissions that may be made. The publisher makes no warranty, express or implied, with respect to the material contained herein.

Printed on acid-free paper

Springer is part of Springer Science+Business Media (www.springer.com)

Preface

Proteomics is widely used in search of biomarkers, in pharmacology, in clinical research and in toxicology. With the help of proteomics, large amount of information on the physiology of living cells can be obtained in a single experiment. Combining this information with the data from genomics and other high-throughput screening techniques, like transcriptomics and metabolomics, allows researchers to gain new insights into the normal and pathological physiology of cells, tissues and organisms.

The area still waiting for the proteomics "boom" is the search for the biological effects of radiation. The effects caused by high doses of radiation, both ionizing and non-ionizing, are relatively well known. However, much less is known about the effects of low or very low doses of radiation to which people are exposed in their everyday life. The knowledge of the effects of the low doses of ionizing radiation (e.g. bystander effect) or low doses of non-ionizing radiation (e.g. effect of radiation emitted by wireless communication devices) is not yet reliably established. Often the effects are small and difficult to discover and to replicate. One of the limiting factors in the research of low-dose radiation effects has been the lack of the knowledge of the cellular target molecules. Now the discovery of molecular targets of radiation is within the reach while using proteomics, the high-throughput screening of expression and activity of proteins. Particularly, in studying biological effects of low-dose radiation, proteomics approach might reveal effects not possible to predict based on the available knowledge of the effects caused by high doses of radiation.

Search through the scientific literature shows that to date there have been published only a very limited number of proteomics studies examining effects of radiation. This book, *Radiation Proteomics*, presents the current status of the research of radiation effects using proteomics approach.

This book begins with the review of current status and the future directions in the development of proteomics methods written by the group of Timothy J. Griffin from the University of Minnesota, Minneapolis, USA. This introductory chapter is followed by two chapters dealing with the effects of ionizing radiation on cells and on tumor microenvironment co-authored by Soile Tapio and Michael J. Atkinson from the Helmholtz Centrum Munich, Germany. Two other chapters dealing with the effects of ionizing radiation present an overview of the radiation effects detectable in the serum proteome, written by Olivier Guipaud from the Institute of Radioprotection and Nuclear Safety in France, and in the urine proteome, co-authored by Mukut Sharma

v

and John E. Moulder from the Medical College of Wisconsin, USA. The next four chapters deal with the effects of non-ionizing radiation. Two other chapters review the impact of electromagnetic fields, the radiofrequency-modulated electromagnetic fields (RF-EMF) and extremely low frequency magnetic fields (ELF-MF), written by Dariusz Leszczynski of the STUK-Radiation and Nuclear Safety Authority, Finland and by Guangdi Chen and Zhengping Xu of the Zhejiang University, China, respectively. The last two chapters by H. Konrad Muller and Gregory M Woods of the University of Tasmania and the Menzies Research Institute Tasmania, Australia and by Riikka Pastila of the STUK-Radiation and Nuclear Safety Authority, Finland, review the effects of non-ionizing ultraviolet radiation on skin cells.

I would like to thank all authors for taking time off their busy schedules to write these interesting reviews. I hope that this book will help to stimulate research in the area of radiation effects on proteome and facilitate discoveries of target molecules, especially in the area of low-dose radiation, both ionizing and non-ionizing.

Finally, I would like to thank Thijs van Vlijmen, Publishing Editor and Sara Germans-Huisman, Publishing Assistant, from the Springer Science + Business Media B.V., for their help support and patiently waiting for the delayed materials.

Melbourne, Australia Dariusz Leszczynski
October 2012

Contents

1 **Mass Spectrometry-Based Proteomics: Basic Principles and Emerging Technologies and Directions** 1
Susan K. Van Riper, Ebbing P. de Jong, John V. Carlis, and Timothy J. Griffin

2 **Ionizing Radiation Effects on Cells, Organelles and Tissues on Proteome Level** 37
Soile Tapio

3 **Radiation Treatment Effects on the Proteome of the Tumour Microenvironment** 49
Michael J. Atkinson

4 **Serum and Plasma Proteomics and Its Possible Use as Detector and Predictor of Radiation Diseases** 61
Olivier Guipaud

5 **The Urine Proteome as a Radiation Biodosimeter** 87
Mukut Sharma and John E. Moulder

6 **Effects of Radiofrequency-Modulated Electromagnetic Fields on Proteome** 101
Dariusz Leszczynski

7 **Global Protein Expression in Response to Extremely Low Frequency Magnetic Fields** 107
Guangdi Chen and Zhengping Xu

8 **Ultraviolet Radiation Effects on the Proteome of Skin Cells** 111
H. Konrad Muller and Gregory M. Woods

9 **Effects of Ultraviolet Radiation on Skin Cell Proteome** 121
Riikka Pastila

Index 129

Mass Spectrometry-Based Proteomics: Basic Principles and Emerging Technologies and Directions

Susan K. Van Riper, Ebbing P. de Jong, John V. Carlis, and Timothy J. Griffin

Abstract

As the main catalytic and structural molecules within living systems, proteins are the most likely biomolecules to be affected by radiation exposure. Proteomics, the comprehensive characterization of proteins within complex biological samples, is therefore a research approach ideally suited to assess the effects of radiation exposure on cells and tissues. For comprehensive characterization of proteomes, an analytical platform capable of quantifying protein abundance, identifying post-translation modifications and revealing members of protein complexes on a system-wide level is necessary. Mass spectrometry (MS), coupled with technologies for sample fractionation and automated data analysis, provides such a versatile and powerful platform. In this chapter we offer a view on the current state of MS-proteomics, and focus on emerging technologies within three areas: (1) New instrumental methods; (2) New computational methods for peptide identification; and (3) Label-free quantification. These emerging technologies should be valuable for researchers seeking to better understand biological effects of radiation on living systems.

Keywords

Proteome • Mass spectrometry • Matrix-assisted laser desorption/ionization • MALDI • Electrospray ionization • ESI • Nanoscale reversed-phase liquid chromatography • NanoLC • Sequence database • SEQUEST • Mascot • 2-dimensional gel electrophoresis • Phosphorylation • Glycosylation • Isotope labeling • Peptide sequencing • Peptide identification

S.K. Van Riper (✉)
Department of Biomedical Informatics and Computational Biology, University of Minnesota, 321 Church St SE/6-155 Jackson Hall, Minneapolis, MN 55455, USA
e-mail: vanr0014@umn.edu

E.P. de Jong
Department of Biochemistry, Molecular Biology and Biophysics, University of Minnesota, 321 Church St SE/6-155 Jackson Hall, Minneapolis, MN 55455, USA
e-mail: dejon039@umn.edu

Abbreviations

2DGE	Two-dimensional gel electrophoresis
APEX	Absolute protein expression
AUC	Area-under-curve
CAD	Collision activated dissociation
CID	Collision induced dissociation
ECD	Electron capture dissociation
ESI	Electrospray ionization
ETD	Electron transfer dissociation
FAIMS	Field-assymetry ion mobility spectrometry
FDR	False discovery rate
HCD	High-energy collision dissociation
HUPO	Human proteome organization
ICAT	Isotope coded affinity tags
IMS	Ion mobility spectrometry
IRMPD	Infrared multiphoton dissociation
iTRAQ	Isotope tagging for relative and absolute quantification
LC	Liquid chromatography
m/z	Mass-to-charge
MALDI	Matrix-assisted laser desorption/ionization
MRM	Multiple reaction monitoring
MS	Mass spectrometry
MS^2	Tandem mass spectrometry
NIST	National institute of standards and testing
NSAF	Normalized spectral abundance factor
PAI	Protein abundance index
PQD	Pulsed Q dissociation
PTM	Post-translational modification
SILAC	Stable isotope labeling of amino acids in cell culture
SRM	Selected reaction monitoring
TMT	Tandem mass tags
Xcorr	Correlation score

J.V. Carlis
Department of Biomedical Informatics and Computational Biology, University of Minnesota, 321 Church St SE/6-155 Jackson Hall, Minneapolis, MN 55455, USA

Department of Computer Science and Engineering, University of Minnesota, 321 Church St SE/6-155 Jackson Hall, Minneapolis, MN 55455, USA
e-mail: carlis@umn.edu

T.J. Griffin
Department of Biomedical Informatics and Computational Biology, University of Minnesota, 321 Church St SE/6-155 Jackson Hall, Minneapolis, MN 55455, USA

Department of Biochemistry, Molecular Biology and Biophysics, University of Minnesota, 321 Church St SE/6-155 Jackson Hall, Minneapolis, MN 55455, USA
e-mail: tgriffin@umn.edu

1.1 Introduction

Genome sequencing efforts initiated in the 1980s fostered a new paradigm in biological research: the system-wide characterization of biomolecules. Within this new paradigm, the field of proteomics, which seeks to characterize proteins on a system-wide level, emerged. Proteins, the major catalytic and structural components within all living systems, are arguably the most informative biomolecules for understanding cellular function and response to systematic perturbations, such as radiation exposure. Unfortunately, proteins are also the most challenging of all biomolecules to study on a system-wide level. In addition to cataloging and quantifying proteins within a complex biological sample, information on their post-translational modification (PTM) state, subcellular localization and interactions with other biomolecules is necessary for full proteome characterization. Adding to the challenge, proteins are dynamic, changing their abundance, PTM state, localization and interactions in response to stimuli. Gene sequences or even mRNA expression levels cannot reveal or predict this protein-level information [1, 2]. Therefore technologies for direct analysis of proteins are necessary for proteome characterization.

Although no single technology can fully characterize all aspects of proteomes, mass spectrometry (MS) is the most powerful and flexible for proteomic analysis. The revolutionary discoveries in the late 1980s of Matrix-Assisted Laser Desorption/Ionization (MALDI) [3] and Electrospray Ionization (ESI) [4] made possible analysis of intact polypeptides and proteins by MS. Along with these ionization methods, three technologies combined to provide an analytical platform underpinning the field of MS-based proteomics and enabling system-wide protein analysis. First, nanoscale reversed-phase liquid chromatography (nanoLC) coupled online with MS instruments came about for separating peptide digests from complex protein mixtures [5]. Second, tandem mass spectrometry, commonly referred to as MS/MS, arose for predictably fragmenting peptides, necessary for determining their amino acid sequence [6]. Tandem mass spectrometry initially scans all mass-to-charge (m/z) values of peptide ions as they elute from the nanoLC column, and records their signal intensities in an MS^1 spectrum. Detected peptide ions are then isolated, and fragmented, with the instrument undertaking another scan of all m/z values of fragment ions, recording their signal intensities in an MS^2 spectrum. Third, automated sequence database searching, led by the program SEQUEST [7] and followed by Mascot [8], was developed to match large amounts of MS^2 spectra to peptide sequences contained in databases, and in turn infer protein identities present within complex mixtures.

This basic platform for what has been termed "shotgun" or "bottom-up" proteomics, offered researchers a new way forward for identifying proteins within complex mixtures. However, two problems, the extreme chemical heterogeneity and large dynamic range of protein abundance within protein mixtures derived from cells, tissues or bodily fluids, required new methods for more sensitive identification of proteins. Multidimensional liquid chromatography-based methods for fractionating peptide digests upstream of MS analysis, helped to, at least in part, address these problems, by simplifying complex mixtures and minimizing signal suppression within the MS instrument [9–11]. These fractionation methods also overcame the limitations [12] of traditionally used two-dimensional gel electrophoresis (2DGE) for separating complex protein mixtures. Methods for enriching PTMs prior to MS analysis improved identification of proteins carrying important modifications, such as phosphorylation [13–15] or glycosylation [16], on a large-scale. Stable isotope labeling and dilution, traditionally used in mass spectrometry analysis of small molecules, was adapted for quantitative measurements of proteins analyzed by MS [17].

Collectively, these components of the MS-based proteomics "toolbox" fostered a new and powerful means to study proteins on a system-wide level. This enhanced platform can now routinely identify and quantify thousands of proteins, including those carrying PTMs in complex protein mixtures. Because proteins are the ubiquitous molecular "effectors" within any organism, MS-based proteomics applies to all fields of biological research, including the effects of radiation on the cellular environment.

MS-based proteomics has always been and remains a collection of dynamic technologies, with new ones constantly emerging across all facets of the platform. Continuous improvements in technologies have moved the proteomics field closer to its ambitious goal to fully characterize proteins within complex biological samples with high throughput. Examples of such technologies include: improvements in MS instrument sensitivity increases identification of low-abundance proteins; more sophisticated software programs for peptide identification from MS^2 data enables detecting a higher proportion of the hundreds of known PTMs [18] of proteins; higher throughput and more quantitatively accurate methods makes possible quantification of protein targets of interest in a large number of individual samples, which is especially important for biomarker studies. However, despite continued technological improvements, the sheer complexity of biological systems greatly challenges the current platform in meeting the goal of full proteome characterization. To illustrate using a rough estimation, the human genome contains about ∼25,000 genes that are processed by a variety of regulated steps (mRNA splicing, proteolysis, etc.) to produce ∼250,000 distinct proteins. These are in turn covalently modified via phosphorylation, acetylation, ubiquitination, oxidation, sumoylation, etc., to generate a

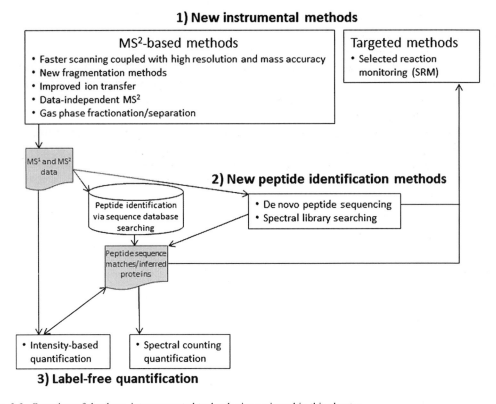

Fig. 1.1 Overview of the three, interconnected technologies reviewed in this chapter

proteome with millions of distinct protein-based molecules. The current proteomics technologies can still only reliably detect a fraction of these molecules. Thus there is continued need for new and improved technologies.

Here, we provide our view on three emerging technologies in MS-based proteomics that are pushing the field in new directions: (1) New instrumental methods; (2) New computational methods for peptide identification; and (3) Label-free quantification. Figure 1.1 provides an overview of the interconnectivity of these technologies. The data produced using new instrumental methods, in particular high resolution and mass accuracy data, enables improved de novo peptide identification, which seeks to overcome the inherent limitations of the currently practiced sequence database searching. Label-free quantification provides a flexible and simple way when comparing samples to determine differentially abundant peptides and inferred proteins. We review recent advances in these three technologies.

1.2 New Instrumental Methods

From the outset, MS instrumentation has been the core technology driving proteomic advances. Fortunately, impressive improvements to the technology have continuously emerged over the last two decades. Most instrument vendors introduce a new model of any given MS instrument every 2–3 years, and those manufactured ~10 years prior to the latest model can scarcely be considered suitable for research. Some of the most fundamental and sought-after metrics for mass spectrometers are resolution, scanning speed, and sensitivity. These are strongly related: in mass spectrometers sensitivity comes at the cost of scanning speed which, in turn, comes at the cost of resolution. Here we review some emerging MS instruments that are redefining what is possible in MS-based proteomic studies. We also discuss emerging methods that are closely linked to improving the performance of the MS instrumentation used for proteomic studies.

1.2.1 Higher Mass Accuracy and Faster Scanning Instruments

Bottom-up proteomics uses nanoLC for peptide separation coupled directly with the MS. Peptides eluting from the nanoLC column are ionized via ESI, and introduced into the mass spectrometer. For complex mixtures (e.g., cell or tissue lysates), the number of peptides vastly exceeds the peak capacity of the separations typically used. Michalski et al. have determined that during a typical 90-min gradient LC run of a complex proteomic mixture, a state-of-the-art mass spectrometer can detect over 100,000 peptide species [19]. Consequently, there are many peptide ions being introduced to the mass spectrometer simultaneously and high resolution is required to differentiate these molecules by their m/z values. Resolution is defined as the ratio of the m/z value to the width of the peak at half its maximum. Therefore a large ratio, for example 50,000/1, is desirable. While not directly related, high mass accuracy usually accompanies high resolution. Mass accuracy is calculated via the following equation: [(actual m/z–observed m/z)/actual m/z]. Because this ratio is usually very small, it is multiplied by 10^6, and reported in units of parts-per-million (ppm). Values of 5 ppm or less are desirable for mass accuracy. Ideally, a mass spectrometer provides sufficient mass accuracy to assign a unique elemental composition, and thus an estimation of amino acid composition, to all peptide peaks in the scanned range. Such a high level of mass accuracy greatly constrains the number of possible amino acid sequences responsible for an observed signal and reduces the incidence of false discoveries when assigning amino acid composition [20]. With 1 ppm measured mass accuracy, the amino acid composition of relatively small peptides with molecular weights in the range of 700–800 Da can be determined [21]. With the help of internal calibration techniques, achieving this level of accuracy now is almost routine [22, 23].

Tandem mass spectrometry is the underlying instrumental analysis method for MS-based proteomics. In its traditional implementation, detected peptide ions eluting from the nanoLC column are isolated and fragmented, with the m/z values of the fragments being recorded in an MS^2 spectrum. There are numerous ways in which isolated peptides can be fragmented, as will be discussed in Sect. 1.4. The most-used method, collision-induced dissociation (CID), leaks or "bleeds" a small quantity of an inert gas (He, N_2, Ar) into the chamber where the isolated peptide ions reside. The peptide ions collide with the gas and internalize the energy from the collision. Being in the gas phase, the peptide ions cannot re-distribute the energy to solvent molecules. Instead, the energy is eventually transferred to a vibrational mode which cannot sustain the energy available and results in bond cleavage. This primarily results in cleavage along the peptide bond of the peptide backbone, although one also frequently sees the loss of water, ammonia, carbon monoxide or labile post-translational modifications [24]. The predominant fragments are named accordingly: b-ions are fragments derived from the N-terminus of the peptide, while y-ions are fragments derived from the C-terminus (see Fig. 1.3 in Sect. 1.2). Certain high-energy fragmentation techniques fragment or completely lose the amino acid side chains, and such ions are named d, v, and w- ions.

Ideally, one would acquire a high-quality MS^2 spectrum for each peptide within a complex mixture. Unfortunately this is not the reality, due to two main factors. First, the speed at which an instrument can gather a sufficient population of peptide ions and generate an MS^2 spectrum will determine its effectiveness at sequencing all the detected peptide ions in a sample. Since the peptide signal from the LC column is transient, the more time spent scanning m/z fragments from any peptide ion selected for fragmentation, the more signals from other peptides will be missed. Thus instruments that quickly scan and record MS^2 spectra are desirable. A second factor is the dynamic range of abundance of the peptides present. The electrospray process can generate only a finite amount of ions per unit time, and when an extremely abundant peptide elutes from the column, less abundant peptides will undergo so-called ion suppression.

Instruments with a greater dynamic range or efficiency at selecting low-abundance ions can mitigate these effects; however peptides from the lowest-abundance proteins in a sample remain undetectable unless enrichment or targeted strategies are employed. Thus, instruments with increased sensitivity to low-abundance peptides are desirable. Increased sensitivity is also linked to scan speed, as increased sensitivity means the instrument must spend less time accumulating fragment ions, and can record MS^2 spectra more rapidly.

Some recently released instruments, combining the desirable qualities of high resolution and mass accuracy and rapid scanning speed, are the Thermo Orbitrap series and the AB Sciex Triple TOF 5600. The Orbitrap mass analyzer allows ions to orbit a central electrode while simultaneously oscillating axially. This axial motion is mass (−to charge) dependent. The Orbitrap analyzer collects an image current of all ions present, each with a characteristic axial frequency. Fourier transform of this image current yields the orbital frequencies present and thus, the m/z values present. In this type of mass analyzer, resolution increases with longer scans [25, 26]. The first commercial Orbitrap instruments, coupled with a linear trapping quadrupole, delivered resolving powers of >100,000 with measured mass accuracy of 2–5 ppm and recording of up to three low-resolution MS^2 scans per second [27]. Recent introduction of the Orbitrap Velos, led to 10 low-resolution MS^2 spectra recorded per second [28]. The latest installment of the Orbitrap series, Orbitrap Elite, employs a more powerful Orbitrap mass analyzer, [29] providing 2–3 fold higher resolution of up to 240,000, and an improved Fourier transform algorithm, delivering a further 2.3-fold greater resolution. No publications using this instrument exist at the time of writing, however a recent publication describes a related instrument. The Q Exactive, employing the same Orbitrap as the Elite but with a detectorless trapping quadrupole, requires that all MS^1 and MS^2 scans be performed in the Orbitrap, thus giving high mass accuracy in all mass spectra and allowing stricter filtering criteria when performing database searches for assignment of peptides to MS^2 spectra. This instrument also records 10 high mass accuracy MS^2 spectra in a ∼1 s cycle that includes an initial MS^1 scan [30]. When coupled to an ultrahigh pressure LC system delivering a 4-h gradient, the Q Exactive achieves 92% coverage of the yeast proteome.

The AB Sciex Triple TOF 5600 is in fact a quadruple time-of-flight (Q-TOF) configuration. Relative to other Q-TOF instruments however, the 5600 has improved ion sampling, rapid pulsing of ions towards the TOF and high TOF acceleration voltages, all of which allow up to 100 MS^2 recorded per second [31]. One of the first publications using this instrument in a proteomics setting determined that 20 MS^2 scans per 1.3 s cycle gave the most peptide assignments to acquired MS^2 spectra. This instrument delivers a resolution of 40,000, and, with internal calibration, also produced a measured mass accuracy of 2 ppm. The extremely fast scanning of this instrument is credited for the threefold increase in peptide identifications over an early-model Orbitrap instrument.

1.2.2 Improved Electrospray Ion Transfer Efficiency

Maximized capture in the mass spectrometer of peptide ions generated via ESI increases the instrument's sensitivity. The ESI process generates a divergent ion beam which is collected by a conductance-limited aperture, typically in the form of a skimmer. This configuration captures only a fraction of the ions generated by ESI. The efficiency of ion transfer from an ESI source to the detector has been estimated at <0.1%. To address this bottleneck, Smith and colleagues have produced many refinements to the long-known stacked ring ion guide [32] or ion funnel (Fig. 1.2), yielding successively improved ion transmission while minimizing the m/z dependency of ion transmission [33–36]. The ion funnel consists of a series of evenly or progressively further-spaced ring electrodes with successively decreasing inner diameters to help focus the divergent ion beam. Radio frequency [32] or static [37] electric fields are used to

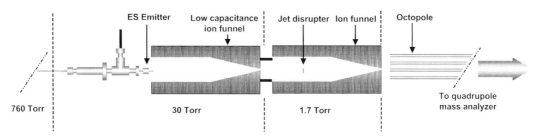

Fig. 1.2 Ion funnel schematic (Adapted with permission [39], Copyright 2008 American Chemical Society)

drive ions through the device sometimes with a DC field superimposed [36]. The ion funnel has achieved collection efficiencies of 50–60% across a typical proteomics m/z range of 200–2,000. However this interface is still not efficiently coupled to a mass spectrometer due to an increased concentration of charged droplets whose repulsion causes losses during transmission [38].

Ion funnels have become widely adopted in commercial mass spectrometers, however the issue of ion transmission remains a barrier to efficient use of all ions generated by ESI. Another factor limiting ion transmission is the pressure gradient between the ESI source and the mass spectrometer. ESI is usually operated at ambient pressure, while the mass spectrometer is operated under high vacuum. This pressure gradient makes efficient ion capture difficult. Operating the electrospray within the low-vacuum region of MS has been used to mitigate this problem and improves ion signal by approximately an order of magnitude [40, 41], giving an estimated 50% ion transmission efficiency [42]. This technology has been termed subambient pressure ionization with nanoelectrospray, or SPIN. Potential, albeit minor hurdles to the widespread adoption of this technique are its compatibility with typical nano-LC flow rates [41] and the robustness of an interface where users are required to introduce a nano-LC-ESI column into the first vacuum stage of a mass spectrometer. However, more efficient use of ions produced by ESI clearly pays dividends in improving instrument sensitivity and will likely continue to see innovations until a majority of ions can be routinely captured in commercial-grade mass spectrometers.

1.2.3 New Fragmentation Methods

Tandem mass spectrometry for amino acid sequence elucidation relies, in part, on the ability to efficiently fragment peptide ions[1]. CID is by far the most used fragmentation method due to its simplicity, ease of implementation, and ability to fragment all peptides at least moderately well, in spite of the wide chemical diversity of a typical proteomic sample. As CID is well-suited to other classes of molecules beside peptides, it is present in virtually all commercially available tandem mass spectrometers [43]. CID is sometimes referred to as collisionally-activated dissociation (CAD), particularly when applied to the beam-type version of this fragmentation. This distinction causes one to view CID as a method limited to resonant excitation in an ion trap. Adding to the confusion, one instrument vendor refers to their CAD cell as Higher energy Collisional Dissociation (HCD) [44]. These distinctions are not merely semantic, however. Aside from the differences in hardware required to perform them, resonant excitation CID is slower (~30 ms vs. <1 ms), produces different fragment ion intensities than beam-type CAD, and CID suffers from the so-called "one-third rule". Under the necessary ion activation conditions for sufficient fragmentation in an ion trap, the resulting fragment ions with m/z ≤ 0.3 times that of the precursor ion are lost during the activation [45]. This loss of low-mass b/y ions as well as helpful immonium ions hinders the interpretation of the resulting tandem mass spectrum.

---

[1] The terms "fragmentation" and "dissociation" are used interchangeably in the field.

In isotope-tagging experiments for relative quantification between samples, the low-mass region of the MS^2 spectra contain the reporter fragment ions, crucial to obtaining quantitative information [46, 47]. While a modified version of resonant excitation in an ion trap, called PQD, can be implemented to preserve the low-mass ions, [48] the low fragmentation efficiency of PQD has invited comparison between PQD and HCD [49]. Instrumental improvements in the HCD cell [50] now make HCD a very attractive method for the analysis of low-mass reporter ions in quantitative mass spectrometry.

The use of HCD fragmentation has also been examined for studies of phosphopeptides. While its use resulted in more phosphopeptide and phosphosite identifications than when using CID, [51] it seems that this improvement can be attributed largely to the high mass accuracy scans which are mandated following HCD, as opposed to low-resolution and low mass accuracy scans typically used following CID, rather than any inherent improvement in fragmentation pattern or ion collection.

Electron capture dissociation (ECD) [52, 53] and electron transfer dissociation (ETD) [54–56] are two related fragmentation methods, used in Ion Cyclotron Resonance (ICR) and ion trap mass spectrometers, respectively. Both methods involve an ion/ion interaction between a multiply protonated peptide cation and either a low-energy electron in ECD or an electron-donating anion radical molecule in ETD. The charge-reduced peptide cation dissociates before any energy randomization can occur. This is especially important for peptides carrying PTMs such as phosphorylation or glycosylation [57]. In CID or CAD the labile covalent bonds between these modifications and the peptide are usually preferentially fragmented, limiting fragmentation across the peptide backbone and resulting in less informative MS^2 spectra for peptide sequence assignment. ECD or ETD meanwhile provide richer MS^2 spectra from many PTM carrying peptides since much of the fragmentation still occurs along the peptide backbone. This leaves intact the modified amino acid residues and also provides a relatively full complement of sequence-rich fragments, enabling more effective sequence assignment and increased confidence in the site of modification. This property has made ECD/ETD especially useful in studies of phosphoproteins and glycoproteins [58–61].

Photodissociation methods can also be used to obtain peptide sequence information. Two spectral regimes are commonly being used for this purpose: infrared and (vacuum) ultraviolet. Infrared multiphoton dissociation (IRMPD) typically uses a CO$_2$ laser emitting tens of watts at a wavelength of 10.6 μm. This wavelength is efficiently absorbed by phosphopeptides and thus IRMPD has been investigated for its utility in analyzing this important post-translational modification [62, 63]. MS^2 spectra following IRMPD do not suffer from the one-third rule, [64] however the fragmentation typically takes twice as long as resonant excitation CID. IRMPD produces b/y ions, but also yields more internal fragment ions than CID [65].

UV photodissociation typically uses excimer lasers emitting at 157 or 193 nm as the light source [66]. As air absorbs these wavelengths efficiently, the 157 nm light source especially must be placed in the vacuum region of the mass spectrometer, complicating the instrumental requirements. Single-photon UV absorption is sufficient to induce dissociation and in contrast to IRMPD, irradiation times on the order of μs or ns are sufficient. While both 157 and 193 nm light target the peptide backbone, UV photodissociation produces a range of fragments in addition to b/y ions such as a, d, x, v and w ions. The presence of d, v and w fragment ions is evidence of a high-energy fragmentation method; not surprising given that the energy of a single UV photon is approximately double that of a peptide bond [67]. While these fragments can be useful, for instance, in differentiating between leucine and isoleucine, most commercially available peptide identification programs are not optimized for, or capable of, analyzing these ions.

While there currently exists an impressive, if not overwhelming, array of dissociation methods, none can meet all the requirements of every conceivable experiment. Until such a method exists, there remains room for improvement to those currently used, and the development of entirely novel ones.

1.2.4 Data-Independent MS² Analysis

Most tandem MS experiments are performed in a data-dependent manner: the collection of peptide ions entering the mass spectrometer are first recorded in the MS^1 spectrum, and these ions (also called precursor ions) are serially selected for fragmentation and MS^2 spectra acquisition [68]. It is well-established however, that this method does not provide complete selection of all peptides in complex samples. An alternative method is to perform fragmentation at all peptide ion m/z values, regardless of which ions can be detected in an MS^1 spectrum. This method is embodied by two different approaches: with and without isolation of precursor ions within a defined m/z window.

In data-dependent MS^2 spectra acquisition, all ions within a defined, relatively narrow m/z window bracketing a precursor of interest are isolated for fragmentation. It however is possible to omit precursor m/z isolation and effectively fragment simultaneously all ions present across the entire m/z range scanned when acquiring MS^1 spectra. When precursor isolation is omitted, a single LC-MS run could in theory detect the entire proteome. In practice, the usual list of mass spectrometer capabilities is desired: high mass accuracy is tremendously beneficial in assigning fragment ions to precursor ions [69, 70]. High scan speed and MS^2 spectra acquisition is beneficial in assigning fragment ions to a precursor, based on chromatographic retention time [71, 72]. Dynamic range of the mass spectrometer is also important in achieving deep sequencing, due to the occurrence of co-eluting peaks [73]. One advantage of this approach is that it can be performed on relatively simple instrumentation: only a collision cell (or other means of achieving dissociation [63]) and a single-stage mass analyzer are required. One major challenge in this type of experiment is the data analysis. Knowledge of which precursor ion masses give rise to the observed fragments is necessary for assigning peptide sequence to MS^2 spectra using sequence database searching software. When simultaneously fragmenting multiple peptide ions across a large m/z range, knowledge of which precursor ion belongs to which fragments is lost. While the precursor mass belonging to sets of fragments can be inferred by relating its retention time in an MS^1 scan to that of the fragments ions in an MS^2 scan, [71] this is not a trivial process [72]. As such, data-independent MS^2 in the absence of precursor isolation still struggles with very complex samples, but this approach seems to be re-evaluated each time a breakthrough in hardware performance is made.

Recently, another data-independent acquisition approach was investigated by rapid isolation and fragmentation of peptide ions within narrow (2.5 m/z) precursor isolation windows, spanning the entire m/z range covered by peptide ions (∼400–1,400). These narrow m/z "bins" mitigated the need for MS^1 scans, while still providing a tight mass range of potential precursor m/z that could be connected to each MS^2 spectra for sequence assignment. For thorough analysis of a typical, complex protein digest, this approach required over 4 days of mass spectrometry instrument time, but required no sample pre-fractionation [74]. Wider isolation widths have been tested, but the resulting tandem mass spectra are likely to contain more than a single peptide species, resulting in complicated database searches [75]. The use of narrow isolation widths demonstrated the ability for a highly automated method to achieve greater proteome coverage and a wider dynamic range than a data-dependent method. As with experiments that do not use precursor isolation, such studies using narrow isolation widths benefit from instrumental improvements such as high mass accuracy and resolution [76]. It is somewhat surprising how few publications exist on this topic, as it seems well-suited to those experimenters not well versed in multidimensional peptide fractionations who might be attracted to a highly automated method. At this time it is difficult to predict whether the data-independent approach will flourish or flounder, in spite of its demonstrated potential.

1.2.5 Gas-Phase Fractionation and Ion Mobility Separations

Since the complexity of a typical proteomics sample can easily exceed the capacity of a LC-MS system to resolve and detect all peptides present, most fractionation schemes [9–11] occur upstream of the mass spectrometer, and are designed to simplify the mixtures introduced into the mass spectrometer to achieve better sensitivity. However, peptide fractionations usually require considerable manual labor and sample handling.

In contrast to upstream fractionation, a fractionation method has been devised wherein repeated injections of the same unfractionated sample are introduced to the LC-MS, but for each injection a different "fraction" of the standard m/z range is analyzed (e.g. 400–575, 560–740, 730–910 and 900–1,795). This allows the instrument to focus on a smaller m/z range to achieve the most comprehensive detection and fragmentation of peptide ions in this range as possible [77–79]. Since the instrument analyzes or ignores certain portions of the ionized m/z range, this method has been termed "gas-phase fractionation". For a yeast cell lysate, the analysis of three gas-phase fractions was compared to triplicate analyses of the entire mass range, and found to increase the number of identifications by 30% [80]. A further refinement of this method used in silico calculations to determine the optimal m/z bins which would yield equal numbers of theoretical tryptic fragments across the number of bins selected [81]. The authors studied three different organisms of differing complexity, and found that regardless of the biological source, roughly half the tryptic peptides reside below m/z 685 with decreasing ion density as m/z increased. Thus gas-phase fractionation certainly has the power to increase proteomic coverage, but at the cost of performing multiple LC-MS runs. Unlike upstream peptide fractionation methods, gas-phase fractionation does this in an entirely automated fashion, reducing labor and sample handling. However, this method might not be suitable to the analysis of very small samples with low protein amounts where multiple LC-MS analyses are not possible.

Ion mobility spectrometry (IMS) is a gas-phase separation method for electrophoretically separating ions in the presence of a buffer gas. Ions are separated by their mass, charge and mobility; the latter being inversely related to their collisional cross section [82]. IMS devices are frequently coupled to a mass spectrometer (using ion funnels), creating a hyphenated method, IMS-MS. For the purposes of this section it will be assumed that all IMS separations are coupled a mass spectrometer. The time frame of a typical IMS separation is ideally suited to its incorporation in a multidimensional fractionation scheme in proteomics: the peak widths for LC, IMS and TOF-MS are on the order of seconds, ms and μs, respectively. This allows each subsequent method to acquire tens of measurements of the preceding separation—the minimum required for adequate profiling of a peak [83].

Three versions of IMS are used: linear drift tubes, traveling wave ion guides, and field-asymmetry IMS (FAIMS) [69]. Linear drift tubes and traveling wave ion guides both resemble a stacked ring ion guide (see Sect. 1.3), though differ in the way electric fields are applied in order to propel ions through the device. These differences affect the separation mechanism. The resolution of linear drift tubes and traveling wave ion guides [84] is typically the greatest at 100–150, however similar values have recently been reported with FAIMS [85, 86]. Also, FAIMS typically separates isomers and isobars better than linear drift tubes and hence has been the most widely implemented in proteomics experiments [85–87]. To date, the most successful configuration for FAIMS is the use of parallel plates separated by ~2 mm [88]. Under high electric fields, the absolute mobility of an ion deviates from its value at low fields. This difference is exploited in FAIMS by applying an asymmetric radio frequency potential between the two plates. As this potential ejects all ions radially from the device, a DC compensation voltage is required to transmit any ions. This compensation voltage is the discriminating variable in a FAIMS separation [87]. Both Thermo Scientific and AB Sciex have commercially-available FAIMS devices which can be added to their mass spectrometers, boasting claims of improved selectivity and signal-to-noise ratios. Shvartsburg, Smith and co-workers have made great improvements in the instrumental design of FAIMS devices, improving resolving power [87, 88] and resolving phosphopeptide isomers which differ only in the site of phosphorylation [89]. Waters Corporation has investigated and commercialized a traveling wave ion guide with its mass spectrometers. As the name implies, a DC voltage is passed along the successive ion guide rings, propelling the ions through the device while an rf-field is generated to maintain ions' radial position. By selecting the amplitude and velocity of the DC wave, ions can be separated by mobility or simply transmitted through the device [90, 91].

Ion mobility separations have the ability to add a second dimension of online fractionation to an LC-MS analysis which should greatly simplify the mixture of ions arriving at the mass spectrometer with no increase in analysis time. The resolving power is sufficient to separate different components in a mixture, however a single peptide sequence may have multiple conformations each with different IMS mobility, which de-focuses the ion packed generated by LC-MS. Also, it is not clear whether current computational methods can analyze IMS separations quickly enough to make on-the-fly decisions, as is currently performed in data-dependent LC-MS experiments. Nonetheless, it seems probable that these hurdles can be overcome and that IMS separations will greatly increase the power of proteomics experiments.

1.2.6 Targeted MS

As the collection of known, MS-observable, proteolytically-derived peptides becomes saturated, some researchers are turning away from data-dependent MS analyses. For a known sample type (e.g. identity of the organism, biological state, sample preparation parameters), the observable peptides emanating from its proteome can be predicted and have likely already been observed in other experiments. Thus, generating a comprehensive list of such so-called "proteotypic" peptides should provide a basis for performing targeted MS experiments in a hypothesis-driven manner [92]. Such methods can then be used, for example, to validate potential biomarkers generated from an initial screening experiment [93] or to follow the proteins in a metabolic pathway following some perturbation [94]. This approach has been greatly advanced by the groups of Aebersold and Carr, who have developed software to predict the most detectable peptides in a mixture, [95, 96] catalogued all experimentally observed peptides [97], demonstrated single copy per cell sensitivity, [94] and are in the process of synthesizing a complete proteotypic peptide library of human serum [98].

A powerful MS method for quantifying several peptides simultaneously is termed selected reaction monitoring (SRM), or sometimes multiple reaction monitoring (MRM). This type of experiment is performed with a triple-quadrupole instrument, and is notoriously selective and sensitive. As ions are electrosprayed into the MS, the first quadrupole transmits a peptide ion at a user-specified m/z value. This ion is then fragmented in the second quadrupole which is not mass-selective, but merely a fragmentation cell. The third quadrupole is then set to transmit the m/z of an expected fragment ion from the precursor peptide ion. This process is repeated, usually for at least three fragments per peptide ion and two proteotypic peptides per protein of interest. Modern mass spectrometers can achieve reliable quantification by dwelling on such a peptide/fragment m/z pair (the "reaction" in SRM, also called a transition) for 10 ms or less. The duty cycle, and thus the sensitivity is inversely related to the number of transitions being monitored, however when the retention time of a peptide is known, the instrument can be scheduled to monitor distinct peptide ions and their transitions at different times. In this way, Kiyonami et al. have quantified 6,000 transitions, relating to 757 peptides in a single LC-MS analysis [99]. They note that this can be extended to 10,000 transitions, targeting 1,000 peptides. Addona et al. [100] have shown that, when using isotopically-labeled standards, this method is very reproducible within and across eight laboratories using two instrument platforms. Many groups believe that SRM-based targeted proteomics will be the basis for future biomarker validation [101–104].

An important aspect of such large-scale hypothesis-driven efforts is the software. The identification of proteotypic peptides and their SRM transitions can be very time-consuming if performed manually. A variety of software products exist from the instrument manufacturers and from academic groups to assist in the design of SRM experiments [105–107]. The most powerful and popular tool has come from the MacCoss laboratory. Their open-source platform, Skyline, can guide SRM experiments by optimizing collision energy and fragment ion selection, performing quantification, predicting peptide retention time and a host of other functions, for data acquired from the major instrument manufacturers [108–110]. Continued refinements to such software packages will greatly automate and thus expedite the process of developing and optimizing SRM assays capable of quantifying hundreds to thousands of peptides in a single MS analysis. These advancements are transforming MS-based proteomics from just a large-scale discovery technology to a high-throughput assay for monitoring proteins of interest in hypothesis-driven studies.

1.3 New Peptide Identification Methods

1.3.1 Principles of MS^2 Fragmentation

Tandem mass spectrometry-based proteomics experiments rely on the same principle as Edman degradation, a long standing chemical technique for peptide sequencing [111]. In Edman degradation stepwise degradation from the peptide's n-terminus followed by chromatographic analysis of the released derivatives determines the amino acid sequence. The fragmentation that occurs during the MS^2 stage mimics Edman degradation because MS^2 dissociation randomly breaks along the backbones between amino acid residues. This results in two, rarely more, fragment ions, one each containing the n-terminus and the c-terminus. The m/z values of fragment ions are recorded in the MS^2 spectra for every selected precursor peptide ion. However, individual fragmentation peaks are not valuable; as in Edman degradation, it is their m/z differences that are informative. As shown in Fig. 1.3, the m/z differences between these peaks determine both the amino acid residue identities and their positions, thus identifying a peptide.

These two fragment ions have predictable structures because as shown in Fig. 1.4, fragmentation can only occur in three places along an ion's backbone. Therefore, the fragment ion will resemble one of six ion structures. The standard nomenclature for these fragment ions identifies both the point of fractionation as well as which terminus retains the charge. Ions a, b and c are n-terminus fragments and x, y and z are c-terminus ions.

Although the exact point of fragmentation depends on many factors, the primary factor is the type of dissociation applied. CID and HCD produce primarily b and y ions, with a few a ions sprinkled in, while ETD produces primarily c and z ions. Their resulting fragmentation patterns differ enough to impact the programs interpreting mass spectra.

1.3.2 Interpretation of MS^2 Spectra

With just one experiment generating hundreds of thousands of MS^1 and MS^2 spectra with high resolution, today's mass spectrometers now offer unparalleled mass accuracy and efficiency. Coupled with the increasing use of new dissociation techniques and chromatography methods, mass spectrometers now generate an overwhelming amount of spectral data with different fragmentation patterns and retention time profiles. Unfortunately, widely used software packages for interpretation of mass spectra, that is, for peptide identification, protein inference and validation, were not designed to process this vast amount of data, and they were tuned to process CID-derived data. Because these tools for interpreting mass spectra have failed to keep pace with advances

1 Mass Spectrometry-Based Proteomics: Basic Principles and Emerging Technologies and Directions 13

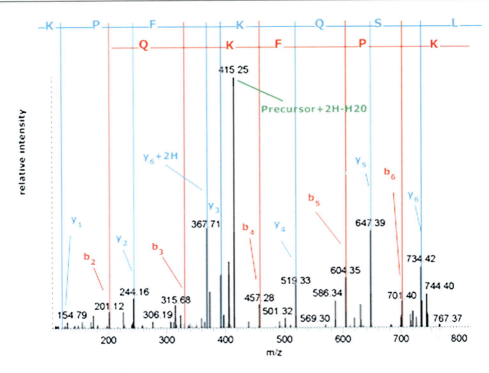

Fig. 1.3 Example peptide MS^2 spectrum

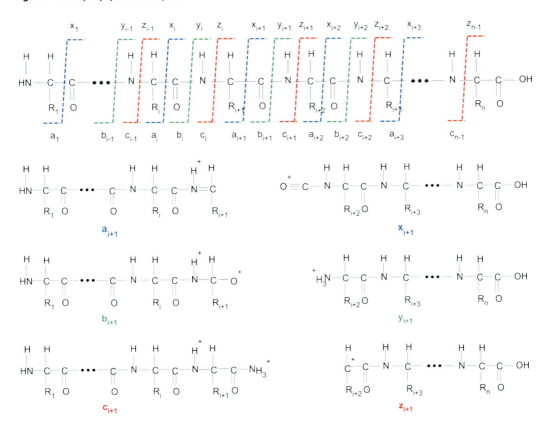

Fig. 1.4 Peptide fragmentation locations and the resulting ions

Table 1.1 Partial list of database search tools

Program	Reference	Website
SEQUEST	[7]	www.thermoscientific.com
Mascot	[8]	www.matrixscience.com/
X!Tandem	[117]	www.thegpm.org/TANDEM/
OMSSA	[118]	www.pubchem.ncbi.nlm.nih.gov/omssa/
Andromeda (MaxQuant)	[119]	www.maxquant.org/
SpectrumMill		www.chem.agilent.com
MyriMatch	[120]	http://fenchurch.mc.vanderbilt.edu/software.php
ProteinProspector	[121]	http://prospector.ucsf.edu/prospector/mshome.htm
PHENYX	[122]	www.genebio.com/products/phenyx

in instrument technology, they yield suboptimal proteome characterization.

Interpretation of mass spectra is a multistep process. The data must first be preprocessed to remove noise and identify valid peaks and features, subjects not reviewed here, but several good reviews exist in the literature [112–115]. After preprocessing the sample data, a series of phases culminates in a list of peptides and/or proteins that are confidently deemed present in the sample. These phases are: peptide identification, protein inference, and validation. In the following sections we highlight their main challenges and solutions, and posit an outlook of their future.

1.3.2.1 Peptide Identification

The first phase, peptide identification assigns an amino acid sequence to a spectrum. This is called a peptide spectrum match or PSM. Peptide identification programs have evolved over time, but strategies for assigning PSMs fall into one of four categories: database search, spectral library search, de novo sequencing, and hybrids thereof.

Database Search

During the early days of proteomics experiments, peptide identification was completed via manual de novo sequencing, a tedious process carried out by researchers without the aid of a computer or a database [116]. However, soon proteomics experiments became high throughput and the amount of data generated by them outpaced researcher's ability to manually inspect each spectrum. This drove the invention of alternate means of identifying peptides, mainly database search programs.

Today, researchers avail themselves to the numerous software packages that implement database search programs, see Table 1.1. Researchers still commonly utilize the first widely used database search programs from the 1990s, SEQUEST [7] and Mascot [8]. Although specific implementations of database search programs differ, they share a common underlying principle introduced by SEQUEST: they compare the observed MS^2 spectra to that of theoretical spectra derived from in-silico enzymatic digestion of a FASTA database. They also share common challenges. One challenge is how to efficiently search the large amount of data available in FASTA databases. Searching all possible peptides from a FASTA database and all of their potential PTMs is prohibitively time-consuming. Even with the use of multiple processors, sequence assignment, including possible PTMs, to hundreds of thousands of MS^2 spectra produced by modern instruments can take days or even weeks. Unfortunately, limiting the peptides only to those with expected enzyme cleavage sites (e.g., lysine and arginine for trypsin cleavage), and limiting the number of PTMs considered, does not adequately narrow the search space. To address this issue, most database search software packages can restrict the search space even further by searching against only those peptides that have a mass within a narrow tolerance window around the observed m/z of its precursor peptide ion. A completely different challenge stems from the fact that different dissociation methods produce very different fragmentation patterns. This was not a problem until recently, because prior to the introduction of ETD, the predominant workhorse of proteomics experiments was CID. But with the introduction of ETD and its increasing adoption comes the requirement for database search programs to allow multiple fragmentation patterns for the same peptide. Because each type of experiment has its own optimal settings for precursor ion mass tolerance window setting, number of PTMs considered and dissociation methods used, the researcher sets these parameters.

An inconvenient consequence of parameter driven database search is that each different set of parameters produces different results. Therefore, researchers must exercise caution when comparing results between experiments, both within and between laboratories.

Although database search strategies are the predominant choice for peptide identification in shotgun proteomics [123], they do have limitations. First, database search relies on the sequencing data for organism being studied. Thus, if an organism is not yet been sequenced, database searching can only be used to find homologous peptides in different organisms. Second, unexpected, yet important, PTMs, and sequence anomalies will be missed because variants do not exist in the database. Even though some databases take into consideration splice variants, no production quality database search engines make the effort to take advantage the annotation available in databases such as Swiss-Prot or UniProtKB. Therefore, many unexpected, but annotated, PTMs and polymorphisms are missed, which leads to incorrect or missed peptide identifications [124]. Third, false positive identifications occur often because database search programs assign a peptide sequence to each and every spectrum, regardless of quality. Fourth, validation assigns high confidence (>95%) to only 10–30% of the spectra [125]. Finally, each database search engine identifies a partially overlapping but different set of peptides. For instance, SEQUEST may identify a set of 100 peptides and Mascot may identify a set of 100 peptides from the same spectra, but perhaps only 60 of them are common to both SEQUEST and Mascot.

A relatively recent development in shotgun proteomics research combines the results from several different database search engines to identify more peptides with increased confidence.

The idea of combining results from multiple sources is not new. Resing et al. described consensus scoring for multiple peptide identifications from different search engines in 2004 [126] and Alves et al. proposed combining and calibrating confidence scores from multiple search engines into a meta-analytic value for each confidence score [127]. However, software automating integration of separate database search results developed more recently. For instance, a popular tool allowing researchers to combine results from multiple search engines is Scaffold, developed by Searle et al. [125, 128]. By probabilistically combining results from multiple search engines, including SEQUEST, X!Tandem, OMSSA, InsPecT and Mascot, Scaffold increases sensitivity a minimum of 20% with each search engine added [129]. As evidenced by the latest publication of a tool combining results from multiple search engines [130], the idea is garnering more attention and we can we can expect this trend of new tools for incorporating multiple search engines to continue into the foreseeable future.

Spectral Library Search

Spectral library search strategies are similar to database search strategies, except the observed MS^2 spectra are compared to collections of experimentally generated spectra rather than hypothetical spectra [131]. These strategies outperform database search strategies in terms of error rates, speed and sensitivity. Using spectral libraries reduces the time spent repeatedly identify the same identifiable peptides by database searching, [132] but can only identify a peptide if it has been previously analyzed by tandem mass spectrometry and its sequence positively identified. A partial list of spectral library search tools is located in Table 1.2.

Libraries of experimental spectra are available from many sources and provide a rich source of spectral data. Spectral libraries for many organisms are stored at the National Institute of Standards and Technology (NIST). Although, the NIST libraries do not target specific PTMs, specialized libraries for specific modifications are available elsewhere, e.g., PhosphoPep [138] for phosphorylation sites in model organisms and

Table 1.2 Partial list of spectral library search tools

Program	Reference	Website
SpectraST	[132, 133]	www.peptideatlas.org/spectrast/
NIST MSPepSearch	N/A	http://peptide.nist.gov/
BiblioSpec	[134]	http://proteome.gs.washington.edu/software/bibliospec/v1.0/documentation/index.html
X!Hunter	[135]	http://www.thegpm.org/hunter/index.html
ProMEX	[136]	http://www.promexdb.org/home.shtml
HMMatch	[137]	Not available

the open source Ub/Ubl spectral library [139] for ubiquitin and ubiqutin-like modifications. In addition, a wealth of spectral data can be downloaded from one of several proteomic data sharing repositories, e.g., PeptideAtlas [140], Pride [141], Peptidome [142], and Tranche (https://trancheproject.org/).

As the amount of publicly available spectral data grows, the hope is that one day spectra for all peptides detectable by MS (at least for well-studied organisms) will be contained and annotated in publicly available spectral reference libraries. However, until these reference libraries are sufficiently complete, spectral library search strategies will continue to be underutilized [143]. In the meantime, data in spectral reference libraries are a rich source of data that could be used for purposes other than identifying peptides. For instance, spectral data could be mined to provide important insight into fragmentation patterns, which could in turn lead to improved database search or de novo sequencing [144] as well as the development of SRM methods to target specific peptides [97].

De Novo Sequencing

The limited ability of database search and spectral search strategies to identify the unexpected, for example, PTMs, polymorphisms and sequence anomalies drives the need for peptide identification programs that can efficiently handle the enormous amounts of data without sacrificing confidence in their results. While conceptually unchanged, researchers are again turning to de novo sequencing as an alternative to database search to accurately and confidently identify peptides. De novo sequencing programs can identify PTMs, polymorphisms and sequence anomalies because they compute directly on spectra to determine the peptide amino acid sequence, process which does not require searching against FASTA databases [145].

De novo sequencing for proteomics has a long and rich history. Spectra were originally sequenced manually, a process which does not scale well. Therefore, as the amount of data from shotgun proteomics grew, researchers turned to computer science for automated de novo sequencing. In the 1980s, several computational algorithms were introduced that helped [146–150] but proved to be terribly slow because they tended to brute force consider all possible amino acid sequences. In 1990, computational algorithms became more efficient when Bartels represented a spectrum as a graph [151]. Although this type of graph is called a spectrum graph by the proteomics community, it should not be confused with spectral graph theory where a graph's spectrum is defined as the set of eigenvalues of a graph's adjacency matrix, nor with a general graph of nodes and edges, where a node does not have a position and an edge can connect any two nodes. In this novel proteomics spectrum graph representation, the vertices represent the spectrum m/z values and two vertices are linked by edges if their mass difference is equivalent to the mass of an amino acid. Figure 1.5 shows a theoretical spectrum graph of the spectrum in Fig. 1.2. Formally, Bartels defined the problem as:

> Given amino acid masses $M = \{m_1, \ldots, m_{20}\}$, spectrum $S = \{s_1, \ldots, s_c\}$, transform it into a spectrum graph $G(V, E)$ such that $V = \{v_1, \ldots, v_c\}$ and $G = \{g_1, \ldots, g_t\}$ such that v represents a single integer m/z and two vertices $v_n, v_q, q \neq n$ and are connected by directed edge e if $|v_q - v_n| \sim m_t$

Although Bartel's approach is now the de facto basis for most de novo peptide sequencing programs, several unresolved issues limited

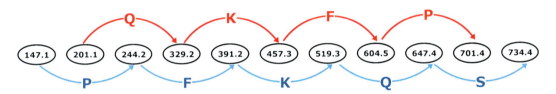

Fig. 1.5 MS^2 spectrum graph

Table 1.3 Partial list of de novo sequencing tools

Program	Reference	Website
PepNovo	[155]	http://proteomics.ucsd.edu/
DirecTag	[157]	http://fenchurch.mc.vanderbilt.edu/software.php
PEAKS	[158]	www.bioinfor.com/
pNovo	[159]	http://proteomics.ucsd.edu/
Lutefisk	[152]	http://www.hairyfatguy.com/lutefisk/
NovoHMM	[156]	http://people.inf.ethz.ch/befische/proteomics/
DeNovoX	N/A	www.thermo.com

spectrum graph's, and, therefore, de novo sequencing's, adoption. First, spectrum graph models were instrument specific which required training a new model for each new mass spectrometer. It took until 1997 to implement this strategy when Taylor and Johnson introduced Lutefisk [152]. Second, the predominantly used CID has a propensity for incomplete fragmentation and results in multiple disconnected graphs, graph gaps, limiting the effectiveness of spectrum graph algorithms. Finally, lack of standardized scoring models for spectrum graphs hindered researchers' ability to compare experimental results. SHERENGA, introduced by Dancik et al. in a landmark publication 1999 [153], addressed each of these limitations using a spectral graph-based algorithm. Several research groups have since made additional enhancements to these algorithms, most notably dynamic programming [154] and probabilistic models using networks learned over annotated spectra (e.g., PepNovo [155] and NovoHMM [156]).

Although several de novo sequencing software packages, as shown in Table 1.3, are now available that implement spectrum graph algorithms, de novo sequencing's adoption as a viable option for shotgun proteomics experiments has been slow. A primary contributor to its slow adoption was that affordable mass spectrometry instrumentation lacked the ability to produce high resolution spectra with minimal noise, completely fragment selected ions and retain potentially important PTMs. Until recently, these issues could only be overcome by using complementary spectra, MS^2 and MS^3 [160], ECD and CID [161], or differentially modified pairs [162] on a FTICR mass spectrometer. The FTICR spectrometer can generate spectra with high resolution (>100,000), thus differentiating valid ions from noise much easier. Furthermore, ECD is complementary to CID, and a more complete fragmentation pattern emerges their spectra are combined into a single artificial spectrum. Finally, ECD inherently uses lower energy than CID allowing for retention and subsequent identification of more PTMs. FTICR mass spectrometers' main drawbacks are that they are extremely expensive and inefficient compared to ion trap mass spectrometers, the main workhorse instruments in MS-based proteomics. Even though FTICR instruments offer 100 times better resolution than an ion trap, each spectrum takes 10 times longer to acquire than on an ion trap instrument.

Recently, via the introduction of the Orbitrap instrument series, the more affordable ion trap mass spectrometers became capable of high resolution (>100,000) and offered ETD, which gives similar spectra to ECD. Taking advantage of these improvements, Datta and Bern expanded on previous pioneering work fusing ECD and CID spectra [163]. In 2009, they introduced Spectrum

Table 1.4 Partial list of hybrid search tools

Program	Reference	Website
InSpecT	[166]	http://proteomics.ucsd.edu/Software/Inspect.html
GutenTag	[165]	http://fields.scripps.edu/
DirecTag	[157]	http://fenchurch.mc.vanderbilt.edu/software.php
TagRecon	[167]	http://fenchurch.mc.vanderbilt.edu/software.php
PeaksDB	[158]	www.bioinfor.com
Paragon (ProteinPilot)	[168]	www.absciex.com

Fusion which uses a global graph partitioning approach to both separate b and y ions and to fuse CID and ETD. The heart of Spectrum Fusion is a supervised machine learning algorithm (tree augmented naïve Bayes network) trained on confidently identified spectra from a prior database search. The result is a synthetic spectrum with only b ions which can then be sequenced by a slightly modified spectrum graph de novo algorithm.

Hybrid Strategies: De Novo & Database/Spectral Library Search

Despite advances in both mass spectrometry instrumentation and software programs, incomplete fragmentation remains an open issue for de novo sequencing strategies. However, when dissociating thousands of ions, they often break along the backbone in enough places so that de novo programs can sequence and identify short peptide sequences which are typically 3–5 short amino acids in length. Again, these ideas are not new. In fact, Mann and Wilm introduced the notion of using short peptides sequences, which they called sequence tags, in 1994 [164], the same year as SEQUEST. However, strategies based on sequence tags did not appear until Tabb et al. published GutenTag program in 2003 [165]. The innovation of GutenTag is that it constructs a model spectrum of the peaks expected from a given sequence tag, compares the observed spectrum and the model spectrum, and generates a correlation score. Tabb et al. went on to provide an enhanced database search tool MyriMatch [120], which is tuned to use these short peptide sequences to infer candidate proteins. Hybrid peptide identification strategies using sequence tags are gaining popularity and several hybrid tools are now available (Table 1.4).

1.3.2.2 Protein Inference

While peptide identification is a necessary phase in proteome profiling, it is not the last one. Proteins must be inferred from the list of peptides identified. However, the task of assembling peptide identifications to infer proteins present in a sample, known as the protein inference problem, is far from trivial [169]. First, the connection between peptides and proteins is lost during enzymatic digestion. This is so because of multiple proteins sharing peptides. The sources of these shared peptides, also known as degenerate peptides, include both natural and artificial phenomena. Degenerate peptides arise often in nature, especially in eukaryotic organisms due to the presence of homologous sequences or splice variants. To make matters worse, errors and redundancies in the database being searched add even more, albeit artificial, degenerate peptides [170]. Regardless of their source, degenerate peptides limit the ability to differentiate between proteins resulting in an unsatisfactory level of ambiguity. This drives the need for validation of results.

1.3.2.3 Validation

MS-based proteomics results are inherently prone to inaccuracies. Without careful filtering, its results are riddled with false positive identifications at both the peptide identification and protein inference levels. To reduce the number of false positives, several scoring models have been proposed and developed to impart a confidence level on identified peptides and inferred proteins. To date, because no single scoring model dominates, different software packages employ their own scoring models. SEQUEST, X!Tandem, Mascot and OMSSA employ variations of a cross correlation (XCorr) score which measures the similarity at different offsets between pre-processed observed spectra and hypothetical spectra generated by in-silico digestion. Mascot differs slightly from other XCorr based scoring models in that it assesses the probability of a peptide spectrum match being a random event. Other software packages use scoring models based on empirically observed rules, SpectrumMill, or incorporate statistically derived fragmentation frequencies, PHENYX [122].

Each of the thousands of single peptide identifications or protein inferences can be assigned an individual score. However, single case scores do not take into consideration the fact that multiple hypotheses are being tested. Therefore, in addition to using a single statistic, p-value, its close relative for multiple testing, E-value, is often used. p-value, assuming the null hypothesis is true, represents the probability of obtaining a test statistic at least as extreme as the one observed. E-value, assuming the null hypothesis is null, is the expected number of times in multiple testing to obtain a test statistic as extreme as the one that was actually observed. Put more simply, E-values are derived by taking the number of tests multiplied by the p-value. To account for multiple hypotheses testing, many controlling measures have been proposed. Bonferroni correction is used to control Family Wise Error Rate, FWER, which is the probability of finding at least one false positive. However, the Bonferroni correction has been shown to be too conservative given the thousands of hypothesis tests in a single experiment [171].

Less conservative than the Bonferroni correction is the False Discovery Rate (FDR) controlling procedure, introduced by Benjamini and Hochberg [172]. They define FDR as the "expected fraction of mistakes among the rejected hypothesis and suggested to control FDR in multiple testing". A well-established mechanism to implement FDR for database search results is to search against a decoy FASTA database of invalid peptide sequences, most often concatenated to the end of the target FASTA database with valid peptide sequences [173]. The premise for this approach is that a spectrum will match valid and random (invalid) sequences with equal probability and target and decoy sequences do not overlap. Although decoy databases are intended to be random, in practice they are most often constructed by reversing, shuffling or randomizing the target FASTA database [174].

With the introduction of FDR as a controlling procedure, publications ensued discussing the proper use of statistical values. Kall et al., argue that using a p-value threshold for FDR is inadequate because the statistical test is performed so many times [175]. It also has the unfortunate property that two different p-value scores can result in the same FDR. To address this problem, Storey and Tibshirani [176] propose a q-score, which when applied to shotgun proteomics, is the defined as the minimum FDR threshold at which a given PSM will be accepted.

Historically, to implement the FDR controlling procedure with a decoy database, researchers accepted all identifications above a certain threshold [177]. This threshold was usually a combination of scores provide by the database search engine. However, problems exist in this strategy, including the need to have separate thresholds for different types of instruments. To overcome problems with the threshold scheme, early validation tools, e.g., QSCORE [178], were developed that employed simple probability, but focused on results from a single search engine.

Because threshold statistical models tend to be instrument specific, researchers turned to machine learning, notably mixture modeling, to build a generic model that could process results from multiple instrument types. Mixture modeling uses models of two normal distributions, one for correct identifications and one for incorrect distributions, to determine a score threshold. Perhaps the most widely used example of mixture modeling for peptide identification validation is Keller et al.'s PeptideProphet [173]. It uses a discriminant score which is derived by converting several scores from the database search programs into a single score. To apply a two-component mixture model, PeptideProphet creates a histogram of discriminant scores and uses curve fitting to draw the correct and incorrect distributions. Using Bayesian statistics, it computes the probability of an identification being correct given its discriminant score. Similar to PeptideProphet is ProteinProphet, which is used to validate protein inferences. It uses results from PeptideProphet as input to accurately compute the probability that an inferred protein is present in the sample [179] and derives a mixture model of correct and incorrect protein inferences, using an expectation-maximization routine (EM). Since PeptideProphet/ProteinProphet is open source and freely available and integrated into the Trans-Proteomic Pipeline, TPP, it is an attractive option for interpreting mass spectra as evidenced by its use in a number of prominent laboratories.

Although scoring based on mixture modeling can accurately model incorrect and correct score distributions, they are inherently complex and not easily extensible [180]. Therefore, other score models have been proposed. For instance, IDPicker is based on a simple non-parametric Monte Carlo simulation method. IDPicker employs FDR identification aggregation instead of individual identification probabilities, and it is easily extended to accept scoring metrics from multiple search engines, as long as the decoys are provided in the searched database [180].

In a recent departure from the canonical target decoy approach, Kim et al., propose MS-GF which uses generating functions and their derivatives without a decoy database [181]. They argue that by using a decoy database, the proteomics community is de facto acknowledging that it has been unable to solve the following Spectrum Matching problem: "Given a spectrum S and a score threshold T for a spectrum-peptide scoring function, find the probability that a random peptide matches the spectrum S with score equal to or larger than T" [181]. This problem assumes certain underlying distributions on which probabilistic calculations can be applied. Ideally, the underlying distributions would be purely theoretical in nature to allow the direct calculation of probability and expectation values. However, the sheer number possible parameters makes modeling the theoretical underlying distribution impractical [182]. Instead, p-values and E-values are calculated using heuristic algorithms working on empirically derived distributions. In contrast to the heuristic algorithms, MS-GF demonstrates that it is possible to compute the precise number of peptides identified in a huge database, solving the Spectrum Matching problem.

1.3.3 Outlook

Although many difficulties exist in thoroughly characterizing a proteome no consensus has been reached by the proteomics community on which the peptide identification, protein inference and validation strategies should be used. This is largely due to the fact that shotgun proteomics is relatively immature and more complex compared to other fields such as genomics. Whereas the genomics community can readily compare results from experiments conducted in different laboratories, the proteomics community has difficulty doing so because reporting of results is not standardized. For instance, some shotgun proteomics researchers will report proteins inferred from a single peptide while others will only report proteins inferred from two or more distinct peptides. If a peptide is shared between multiple proteins, some researchers randomly assign the peptide to a protein, while others apply Occam's razor or other statistical models. This is compounded by the availability of vastly different FASTA databases for a single organism. The differences mainly stem from their curation processes, or lack thereof, and their sources of deposited sequences.

Reporting standards for shotgun proteomics experiments may be lacking consensus, but serious effort has been made to rectify this problem. In 2002, the Human Proteome Organization, HUPO, launched the Proteomics Standards Initiative, PSI. Its goal was and is to "define community standards for data representation in proteomics to facilitate systematic data capture, comparison, exchange and verification." [183–187]. Although HUPO sets standards for the broader proteomics community, publishing criteria was still lacking for shotgun proteomics results. To address this, about 30 key people in the proteomics community met in Paris to develop set of standards focused on publication of shotgun proteomics results. These standards published in 2006 [188] as the Paris Guidelines, and updated in 2009 are slowly being adopted by proteomics journals.

1.4 Label-Free Quantification

The initial application of the MS-based proteomics platform addressed the challenge of cataloging proteins within complex samples. However, biological researchers also need to quantify proteins because proteomes are highly dynamic systems, and their abundances change due to regulation of their synthesis and degradation. Protein activities are dynamically regulated via the addition or removal of PTMs. Therefore, to make MS-proteomics a technology truly useful to researchers who are trying to understand living systems, it must be able to quantify abundance and PTM differences between samples.

The initial technology for quantitative MS-based proteomics involved differential labeling methods with stable isotopes. Isotope labeling methods for quantitative MS-based proteomics have been reviewed in detail [189, 190]. These methods label proteins and/or peptides with stable isotopes (15N, 13C, 18O) through a variety of mechanisms. Stable isotope labeling in cell culture (SILAC) labels proteins via metabolic incorporation of stable-isotope containing amino acids contained in cell culture media [191, 192]. Other methods introduce stable isotopes via reactive chemical tags, such as the isotope affinity tag (ICAT) [193] or isobaric peptide tagging (e.g. iTRAQ, TMT) [194, 195] methods. Labeling with O^{18} is accomplished via enzymatic means at the c-terminus of peptides within complex mixtures [196]. For all of these methods, distinct protein mixtures are first differentially labeled, one with isotopically normal amino acids or chemical tags, and the other with isotopically "heavy" amino acids or chemical tags. Although most labeling methods compare protein abundance between two distinct mixtures, some are capable of multiplexed analysis, such as iTRAQ labeling which can compare up to eight [197]. After labeling, the mixtures are combined and peptide digests are fractionated and analyzed by MS. Peptide sequences common to both samples, although differentially isotopically labeled, retain the same chemical properties and behave similarly during fractionation. Consequently, differentially isotopically labeled peptides are detected simultaneously and their m/z differences resolved in the MS. Peptides are selected for MS^2 and identified via subsequent sequence database searching. For identified peptides, relative abundance levels between samples are determined via comparison of the mass spectral peak intensities corresponding to the normal or heavy isotope labels.

Although still used prominently, stable isotope labeling has its limitations. One is cost. Stable isotope labeled amino acids or chemical tags are costly to synthesize, and purchase of these can run from hundreds to thousands of dollars, depending on the labeling method used. Another is applicability to only certain biological sample types. SILAC, arguably the most accurate stable isotope labeling method, is only applicable to experiments using cell culture models, although extremely expensive studies of whole organism labeling with stable isotopes in mice and worm have been described [198]. For human and other animal studies, chemical tagging methods, such as iTRAQ or TMT, must be used for stable isotope labeling. Unfortunately, the accuracy of iTRAQ and TMT for measuring relative abundances, which are based on MS^2 fragmentation of labeled peptides, is decreased due to simultaneous fragmentation of multiple peptides in shotgun proteomics [199].

Responding to these limitations, label-free technology has emerged which obviates the need for stable isotope labeling for quantitative proteomics. Two methods underpin the label-free MS-based quantitative proteomics technology: spectral counting and intensity-based measurements. Figure 1.6 details these two methods.

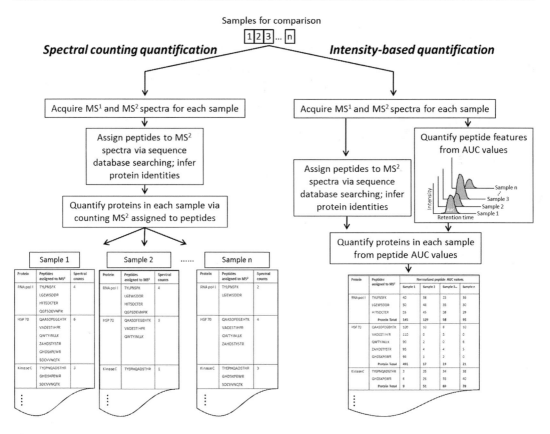

Fig. 1.6 Label-free quantification methods

1.4.1 Spectral Counting Quantification

Spectral counting is based on the core instrumental method used in MS-based shotgun proteomics. Here, peptides separated via LC are detected and selected for CID fragmentation using a data-dependent routine. The fragmentation spectra are recorded as MS^2 spectra. Peptides are identified by assigning a sequence to each MS^2 from databases of known protein sequences and a variety of software programs, as described in Sect. 1.2. Protein identities in the starting mixture are inferred from the identification of peptides that are a part of their amino acid sequence. Quantification via spectral counting is based on the observation that the number of peptides identified from MS^2 spectra is proportional to the abundance of the protein in the starting mixture: more abundant proteins result in more identified peptides while less abundant proteins result in fewer identified peptides. Protein quantification is achieved by simply counting the number of MS^2 spectra assigned to peptides within a given protein, without taking into consideration the peptide MS signal intensity. Because quantification is based on peptides assigned to MS^2 spectra, spectral counting benefits from MS instruments with higher mass accuracy and sensitivity, which increase the number of high confidence peptide identifications [20].

Early on, spectral counting was done in a rather simple manner, simply summing the number of peptide identifications corresponding to each inferred protein. However, as this method increased in popularity, more sophisticated quantification approaches based on spectral counting have emerged. Several extensive reviews have recently appeared on spectral counting [200]. Here we discuss the most commonly used approaches to spectral counting quantification and some representative studies which have used this method.

Table 1.5 Summary of open-source software for label-free quantification

Program	Reference	Website
Spectral counting quantification		
APEX	[205]	http://pfgrc.jcvi.org/index.php/bioinformatics/apex.html/
Census	[210]	http://fields.scripps.edu/census/
emPAI	[202]	http://empai.iab.keio.ac.jp/
PepC	[211]	http://sashimi.svn.sourceforge.net/viewvc/sashimi/trunk/trans_proteomic_pipeline/src/Quantitation/Pepc
SI_N^a	[212]	N/A
Intensity-based quantification		
IDEAL-Q	[213]	http://ms.iis.sinica.edu.tw/IDEAL-Q/
MaxQuant	[214]	http://maxquant.org/
MSInspect	[215]	http://proteomics.fhcrc.org/CPL/msinspect/index.html
MZMine	[216]	http://mzmine.sourceforge.net/download.shtml
PEPPeR	[217]	http://www.broadinstitute.org/cancer/software/genepattern/desc/proteomics#pepper
SuperHirn	[218]	http://prottools.ethz.ch/muellelu/web/SuperHirn.php

[a] Approach combines both spectral counting and intensity-based quantification

Spectral counting must take into account a protein's length because a longer protein, when enzymatically digested, will produce more peptides than a shorter protein for the MS to detect. Without correction, protein quantification by spectral counting would be biased towards longer proteins. As a consequence, an approach taking into account protein length was developed [201] which provided a normalized spectral abundance factor (NSAF) for each identified protein. The abundance of any given protein within a mixture can be estimated by dividing its NSAF value against the sum of NSAF values for all identified proteins.

An alternative approach to NSAF is the protein abundance index (PAI) [202], which was further improved to the exponential modified PAI, or emPAI [203]. This approach used the number of peptides actually identified from a protein, divided by the estimated total number of peptides expected to be identified for that same protein. The expected peptides were estimated based on the proteins sequence and the sizes of peptides derived from the protein after enzymatic digestion. The relative molar amount of any given protein within a sample can then be calculated by dividing its emPAI value against the sum of all emPAI values within the mixture. The emPAI approach was deployed in a freely available application, emPAI Calc that accepts data from a variety of sequence database searching programs [200].

Another approach, Absolute Protein Expression (APEX), tries to correct for physiochemical variations between peptide sequences that may affect their identification in the MS, and bias spectral counting results. APEX uses a correction factor that attempts to use properties such as amino acid content and length of peptides [204] to assess the probability of any given peptide for MS detection and subsequent identification from MS^2 spectra. This correction is applied to the spectral counts corresponding to each identified protein, to provide a more accurate measurement of its abundance. APEX has been released as an open source application [205].

Spectral counting has been widely applied. Its application is reviewed in detail elsewhere [200, 206]. Software plays a key role in the automating spectral counting quantification. Table 1.5 shows a summary of the most popular open-source software available. One particularly powerful application uses spectral counting and NSAF values to quantify relative abundance of proteins within functional complexes. Estimation of relative stoichiometry of the different members of protein complexes [207], as well as modeling of protein-protein interaction networks [201] is possible. An interesting application using the emPAI approach identified and quantified relative abundance levels of over 100 proteins in the chicken egg white proteome [208]. APEX was recently used to characterize proteome abundance differences between mutant strains of the thermophilic anaerobic bacterium *Clostridium thermocellum*, an organism with promise for biofuel production [209].

1.4.2 Intensity-Based Quantification

An alternative to spectral counting is intensity-based measurements of peptide abundance. During a nanoLC-MS analysis, the mass-to-charge (m/z), retention time and signal intensity values are continuously recorded for each detected peptide. This information can be used to reconstruct a chromatographic peak for each peptide. This quantification method estimates the area under the curve (AUC) of the chromatographic peak (Fig. 1.5). The AUC correlates linearly with peptide concentration across a range of low femtomole amounts to tens of picomoles in most contemporary MS instruments [219, 220]. Similar to spectral counting, peptides are identified via MS^2 and sequence database searching, and protein identities are inferred from these peptides. For comparisons of peptide and inferred protein abundance between different samples, each sample is analyzed by nanoLC-MS separately. AUC values calculated for detected peptides in each distinct sample are compared to determine relative abundance. Intensity-based measurements are not used for quantification of different peptides within the same sample, because each peptide sequence ionizes with different efficiency, making comparison based on signal intensity inaccurate.

Although simple in concept, successful implementation of intensity-based quantification relies heavily on sophisticated software. Open-source software choices have been reviewed elsewhere [221]. Some of these choices are summarized in Table 1.5. This software automates critical data processing steps needed to insure accurate results based on AUC values. A recent review by Christin and colleagues [221] thoroughly describes these steps. One key step is proper alignment of peaks corresponding to the same peptide across all separate nanoLC-MS data sets. Proper alignment, based on peak m/z values and retention time, assures that the AUC values being measured in each sample correspond to the same detected peptide. Use of highly reproducible nanoLC systems with high chromatographic resolving power can help for alignment [222], although ultimately effective alignment via software is critical. High accuracy measurements of peptide m/z values using newer MS instruments has greatly helped with alignment across separate nanoLC-MS datasets. One nice feature of peak alignment, aided by high mass accuracy data, is that a peptide need only be identified by MS^2 in one sample [221]. Peaks in other samples aligning in retention time and accurate m/z can then be confidently assigned to that peptide without the need for their identification from MS^2 spectra.

Another key step is normalization of measured AUC values. Normalization accounts for bias and variability in measured AUC values introduced during sample processing, loading of sample to the nanoLC column, and in-run variability of MS response. A number of normalization procedures have been developed which are effective for minimizing variability and improving accuracy [223, 224].

As with spectral counting, applications of intensity-based quantification are numerous. These are reviewed in detail elsewhere [225, 226]. These different applications have used a variety of publically available software programs for accurate quantification, some of which are summarized in Table 1.5. Here we discuss several representative applications. One interesting, radiation research-relevant example, demonstrated the effectiveness of intensity-based quantification to compare effects of ionizing radiation on colon cancer cells compared to a mock-treated control [227]. Disease biomarker discovery has also been a popular application of intensity-based quantification. Such studies have been done in paraffin embedded archival cancer tissues [228], as well as serum fluid from schizophrenia patients [229].

Overall, label-free quantification addresses many of the limitations of stable-isotope labeling-based technology. Both spectral counting and intensity-based measurements are cheap and simple, with no need for purchase of costly labeling reagents or extra sample labeling and processing steps. Spectral counting provides the additional benefit of measuring relative abundance of proteins within the same sample, whereas stable isotope labeling only measures relative abundance across separate samples. Intensity-based measurements, when using effective software for aligning peptide peaks across samples, obviates the need for time- and computation-intensive MS^2 acquisition and subsequent peptide identification via sequence database searching. This method therefore is an attractive choice for biomarker studies, where comparison across many patient samples with high throughput is desirable.

Despite numerous strengths, label-free quantification is not without limitations. Unlike some labeling methods, notably the iTRAQ or TMT methods, multiplexed comparative analysis within a single MS experiment is not possible. Instead, each sample being compared must be analyzed in a separate MS experiment, and preferably with technical replicates to achieve statistical significance [230]. Consequently, large amounts of instrument time are required, which may not be feasible, especially for researchers relying on sample analysis via a fee-for-service facility. Low-abundance proteins also remain a challenge for both methods. Because spectral counting relies on multiple peptides to be identified from each inferred protein to achieve statistical significance, low abundance proteins identified by only a few peptides cannot be accurately quantified. For intensity-based measurements, peptide peaks from low-abundance proteins also suffer from low signal-to-noise ratios, challenging their accurate quantification. Improved instrument sensitivity should only help to increase the ability to identify more peptides derived from low abundance proteins, and improve the effectives of both label-free methods. Recently, a promising new method was described [212] which combines spectral counting and intensity-based measurements, thereby capitalizing on the strengths of both methods and providing improved results.

1.5 Conclusions

Consistent with history, technological advances will continue to define and mature the field of MS-based proteomics, catalyzing new milestones of achievement. We anticipate these advances to primarily fall in the areas described in this review: new instrumentation and related methods, and new computational methods and software for identification and quantification of proteins from complex datasets. Continued maturation of MS-based proteomics should one day enable realization of its ultimate goal: comprehensive proteome characterization. Researchers seeking to better understand the effects of radiation on living systems will undoubtedly continue to benefit from the continued advances of this vital technology.

References

1. Griffin TJ, Gygi SP, Ideker T, Rist B, Eng J, Hood L, Aebersold R (2002) Complementary profiling of gene expression at the transcriptome and proteome levels in *Saccharomyces cerevisiae*. Mol Cell Proteomics 1(4):323–333
2. Washburn MP, Koller A, Oshiro G, Ulaszek RR, Plouffe D, Deciu C, Winzeler E, Yates JR 3rd (2003) Protein pathway and complex clustering of correlated mRNA and protein expression analyses in *Saccharomyces cerevisiae*. Proc Natl Acad Sci U S A 100(6):3107–3112. doi:10.1073/pnas.0634629100 0634629100 [pii]
3. Tanaka K, Waki H, Ido Y, Akita S, Yoshida Y, Yoshida T, Matsuo T (1988) Protein and polymer analyses up to m/z 100,000 by laser ionization time-of-flight mass spectrometry. Rapid Comm Mass Spectrom 2(8):151–153
4. Fenn JB, Mann M, Meng CK, Wong SF, Whitehouse CM (1989) Electrospray ionization for mass spectrometry of large biomolecules. Science 246(4926):64–71
5. Deterding LJ, Moseley MA, Tomer KB, Jorgenson JW (1991) Nanoscale separations combined with tandem mass spectrometry. J Chromatogr 554(1–2):73–82
6. Hunt DF, Yates JR 3rd, Shabanowitz J, Winston S, Hauer CR (1986) Protein sequencing by tandem mass spectrometry. Proc Natl Acad Sci U S A 83(17):6233–6237
7. Eng JK, McCormack AL, Yates JRI (1994) An approach to correlate tandem mass spectral data of peptides with amino acid sequences in a protein database. J Am Soc Mass Spectrom 5:976–989

8. Perkins DN, Pappin DJ, Creasy DM, Cottrell JS (1999) Probability-based protein identification by searching sequence databases using mass spectrometry data. Electrophoresis 20(18):3551–3567. doi:10.1002/(SICI)1522-2683(19991201)20:18<3551::AID-ELPS3551>3.0.CO;2-2 [pii] 10.1002/(SICI)1522-2683(19991201)20:18<3551::AID-ELPS3551>3.0.CO;2-2

9. Gygi SP, Rist B, Griffin TJ, Eng J, Aebersold R (2002) Proteome analysis of low-abundance proteins using multidimensional chromatography and isotope-coded affinity tags. J Proteome Res 1(1): 47–54

10. Link AJ, Eng J, Schieltz DM, Carmack E, Mize GJ, Morris DR, Garvik BM, Yates JR 3rd (1999) Direct analysis of protein complexes using mass spectrometry. Nat Biotechnol 17(7):676–682

11. Washburn MP, Wolters D, Yates JR 3rd (2001) Large-scale analysis of the yeast proteome by multidimensional protein identification technology. Nat Biotechnol 19(3):242–247

12. Gygi SP, Corthals GL, Zhang Y, Rochon Y, Aebersold R (2000) Evaluation of two-dimensional gel electrophoresis-based proteome analysis technology. Proc Natl Acad Sci U S A 97(17): 9390–9395

13. Ficarro SB, McCleland ML, Stukenberg PT, Burke DJ, Ross MM, Shabanowitz J, Hunt DF, White FM (2002) Phosphoproteome analysis by mass spectrometry and its application to *Saccharomyces cerevisiae*. Nat Biotechnol 20(3):301–305

14. Oda Y, Nagasu T, Chait BT (2001) Enrichment analysis of phosphorylated proteins as a tool for probing the phosphoproteome. Nat Biotechnol 19(4): 379–382

15. Zhou H, Watts JD, Aebersold R (2001) A systematic approach to the analysis of protein phosphorylation. Nat Biotechnol 19(4):375–378

16. Zhang H, Li XJ, Martin DB, Aebersold R (2003) Identification and quantification of N-linked glycoproteins using hydrazide chemistry, stable isotope labeling and mass spectrometry. Nat Biotechnol 21(6):660–666

17. Flory MR, Griffin TJ, Martin D, Aebersold R (2002) Advances in quantitative proteomics using stable isotope tags. Trends Biotechnol 20(12 Suppl): S23–S29

18. Creasy DM, Cottrell JS (2004) Unimod: protein modifications for mass spectrometry. Proteomics 4(6):1534–1536. doi:10.1002/pmic.200300744

19. Michalski A, Cox J, Mann M (2011) More than 100,000 detectable peptide species elute in single shotgun proteomics runs but the majority is inaccessible to data-dependent LC-MS/MS. J Proteome Res 10(4):1785–1793. doi:10.1021/pr101060v

20. Mann M, Kelleher NL (2008) Precision proteomics: the case for high resolution and high mass accuracy. Proc Natl Acad Sci U S A 105(47):18132–18138. doi:10.1073/pnas.0800788105

21. Zubarev RA, Hakansson P, Sundqvist B (1996) Accuracy requirements for peptide characterization by monoisotopic molecular mass measurements. Anal Chem 68(22):4060–4063. doi:10.1021/ac9604651

22. Olsen JV, de Godoy LMF, Li GQ, Macek B, Mortensen P, Pesch R, Makarov A, Lange O, Horning S, Mann M (2005) Parts per million mass accuracy on an orbitrap mass spectrometer via lock mass injection into a C-trap. Mol Cell Proteomics 4(12):2010–2021. doi:10.1074/mcp.T500030-MCP200

23. Zhang Y, Wen Z, Washburn MP, Florens LA (2011) Improving proteomics mass accuracy by dynamic offline lock mass. Anal Chem. doi:10.1021/ac201867h

24. Papayannopoulos IA (1995) The interpretation of collision-induced dissociation tandem mass-spectra of peptides. Mass Spectrom Rev 14(1):49–73. doi:10.1002/mas.1280140104

25. Hardman M, Makarov AA (2003) Interfacing the orbitrap mass analyzer to an electrospray ion source. Anal Chem 75(7):1699–1705. doi:10.1021/ac0258047

26. Makarov A (2000) Electrostatic axially harmonic orbital trapping: a high-performance technique of mass analysis. Anal Chem 72(6):1156–1162. doi:10.1021/ac991131p

27. Makarov A, Denisov E, Kholomeev A, Baischun W, Lange O, Strupat K, Horning S (2006) Performance evaluation of a hybrid linear ion trap/orbitrap mass spectrometer. Anal Chem 78(7):2113–2120. doi:10.1021/ac0518811

28. Olsen JV, Schwartz JC, Griep-Raming J, Nielsen ML, Damoc E, Denisov E, Lange O, Remes P, Taylor D, Splendore M, Wouters ER, Senko M, Makarov A, Mann M, Horning S (2009) A dual pressure linear Ion trap orbitrap instrument with very high sequencing speed. Mol Cell Proteomics 8(12):2759–2769. doi:10.1074/mcp.M900375-MCP200

29. Makarov A, Denisov E, Lange O (2009) Performance evaluation of a high-field orbitrap mass analyzer. J Am Soc Mass Spectrom 20(8):1391–1396. doi:10.1016/j.jasms.2009.01.005

30. Michalski A, Damoc E, Hauschild JP, Lange O, Wieghaus A, Makarov A, Nagaraj N, Cox J, Mann M, Horning S (2011) Mass spectrometry-based proteomics using Q exactive, a high-performance benchtop quadrupole orbitrap mass spectrometer. Mol Cell Proteomics 10(9). doi:10.1074/mcp.M111.011015

31. Andrews GL, Simons BL, Young JB, Hawkridge AM, Muddiman DC (2011) Performance characteristics of a New hybrid quadrupole time-of-flight tandem mass spectrometer (TripleTOF 5600). Anal Chem 83(13):5442–5446. doi:10.1021/ac200812d

32. Bahr R, Gerlich D, Teloy E (1969) Verhandl DPG (VI) 4:343

33. Page JS, Tolmachev AV, Tang KQ, Smith RD (2006) Theoretical and experimental evaluation of the low m/z transmission of an electrodynamic ion funnel. J Am Soc Mass Spectrom 17(4):586–592. doi:10.1016/j.jasms.2005.12.013

34. Kim T, Tolmachev AV, Harkewicz R, Prior DC, Anderson G, Udseth HR, Smith RD, Bailey TH, Rakov S, Futrell JH (2000) Design and implementation of a new electrodynamic ion funnel. Anal Chem 72(10):2247–2255. doi:10.1021/ac991412x
35. Shaffer SA, Tang KQ, Anderson GA, Prior DC, Udseth HR, Smith RD (1997) A novel ion funnel for focusing ions at elevated pressure using electrospray ionization mass spectrometry. Rapid Commun Mass Spectrom 11(16):1813–1817
36. Kelly RT, Tolmachev AV, Page JS, Tang KQ, Smith RD (2010) The ion funnel: theory, implementations, and applications. Mass Spectrom Rev 29(2):294–312. doi:10.1002/mas.20232
37. Guan SH, Marshall AG (1996) Stacked-ring electrostatic ion guide. J Am Soc Mass Spectrom 7(1):101–106. doi:10.1016/1044-0305(95)00605-2
38. Kelly RT, Page JS, Marginean I, Tang KQ, Smith RD (2008) Nanoelectrospray emitter arrays providing interemitter electric field uniformity. Anal Chem 80(14):5660–5665. doi:10.1021/ac800508q
39. Page JS, Tang K, Kelly RT, Smith RD (2008) Subambient pressure ionization with nanoelectrospray source and interface for improved sensitivity in mass spectrometry. Anal Chem 80(5):1800–1805. doi:10.1021/ac702354b
40. Page JS, Kelly RT, Tang K, Smith RD (2007) Ionization and transmission efficiency in an electrospray ionization-mass spectrometry interface. J Am Soc Mass Spectrom 18(9):1582–1590. doi:10.1016/j.jasms.2007.05.018
41. Tang KQ, Page JS, Marginean I, Kelly RT, Smith RD (2011) Improving liquid chromatography-mass spectrometry sensitivity using a subambient pressure ionization with nanoelectrospray (SPIN) interface. J Am Soc Mass Spectrom 22(8):1318–1325. doi:10.1007/s13361-011-0135-7
42. Marginean I, Page JS, Tolmachev AV, Tang KQ, Smith RD (2010) Achieving 50% ionization efficiency in subambient pressure ionization with nanoelectrospray. Anal Chem 82(22):9344–9349. doi:10.1021/ac1019123
43. McLuckey SA, Mentinova M (2011) Ion/neutral, ion/electron, ion/photon, and ion/Ion interactions in tandem mass spectrometry: do we need them all? Are they enough? J Am Soc Mass Spectrom 22(1):3–12. doi:10.1007/s13361-010-0004-9
44. McAlister GC, Phanstiel D, Wenger CD, Lee MV, Coon JJ (2010) Analysis of tandem mass spectra by FTMS for improved large-scale proteomics with superior protein quantification. Anal Chem 82(1):316–322. doi:10.1021/ac902005s
45. Louris JN, Cooks RG, Syka JEP, Kelley PE, Stafford GC, Todd JFJ (1987) Instrumentation, applications, and energy deposition in quadrupole ion-trap tandem mass-spectrometry. Anal Chem 59(13):1677–1685. doi:10.1021/ac00140a021
46. Ross PL, Huang YLN, Marchese JN, Williamson B, Parker K, Hattan S, Khainovski N, Pillai S, Dey S, Daniels S, Purkayastha S, Juhasz P, Martin S, Bartlet-Jones M, He F, Jacobson A, Pappin DJ (2004) Multiplexed protein quantitation in *Saccharomyces cerevisiae* using amine-reactive isobaric tagging reagents. Mol Cell Proteomics 3(12):1154–1169. doi:10.1074/mcp.M400129-MCP200
47. Thompson A, Schafer J, Kuhn K, Kienle S, Schwarz J, Schmidt G, Neumann T, Hamon C (2003) Tandem mass tags: a novel quantification strategy for comparative analysis of complex protein mixtures by MS/MS. Anal Chem 75(8):1895–1904. doi:10.1021/ac0262560
48. Griffin TJ, Xie HW, Bandhakavi S, Popko J, Mohan A, Carlis JV, Higgins L (2007) iTRAQ reagent-based quantitative proteomic analysis on a linear ion trap mass spectrometer. J Proteome Res 6(11):4200–4209. doi:10.1021/pr070291b
49. Bantscheff M, Boesche M, Eberhard D, Matthieson T, Sweetman G, Kuster B (2008) Robust and sensitive iTRAQ quantification on an LTQ orbitrap mass spectrometer. Mol Cell Proteomics 7(9):1702–1713. doi:10.1074/mcp.M800029-MCP200
50. Pichler P, Kocher T, Holzmann J, Mohring T, Ammerer G, Mechtler K (2011) Improved precision of iTRAQ and TMT quantification by an axial extraction field in an orbitrap HCD cell. Anal Chem 83(4):1469–1474. doi:10.1021/ac102265w
51. Nagaraj N, D'Souza RCJ, Cox J, Olsen JV, Mann M (2010) Feasibility of large-scale phosphoproteomics with higher energy collisional dissociation fragmentation. J Proteome Res 9(12):6786–6794. doi:10.1021/pr100637q
52. Zubarev RA, Kelleher NL, McLafferty FW (1998) Electron capture dissociation of multiply charged protein cations. A nonergodic process. J Am Chem Soc 120(13):3265–3266. doi:10.1021/ja973478k
53. Bakhtiar R, Guan ZQ (2006) Electron capture dissociation mass spectrometry in characterization of peptides and proteins. Biotechnol Lett 28(14):1047–1059. doi:10.1007/s10529-006-9065-z
54. Coon JJ, Ueberheide B, Syka JEP, Dryhurst DD, Ausio J, Shabanowitz J, Hunt DF (2005) Protein identification using sequential ion/ion reactions and tandem mass spectrometry. Proc Natl Acad Sci U S A 102(27):9463–9468. doi:10.1073/pnas.0503189102
55. Syka JEP, Coon JJ, Schroeder MJ, Shabanowitz J, Hunt DF (2004) Peptide and protein sequence analysis by electron transfer dissociation mass spectrometry. Proc Natl Acad Sci U S A 101(26):9528–9533. doi:10.1073/pnas.0402700101
56. Mikesh LM, Ueberheide B, Chi A, Coon JJ, Syka JEP, Shabanowitz J, Hunt DF (2006) The utility of ETD mass spectrometry in proteomic analysis. BBA-Proteins Proteomics 1764(12):1811–1822. doi:10.1016/j.bbapap.2006.10.003
57. Zubarev RA (2003) Reactions of polypeptide ions with electrons in the gas phase. Mass Spectrom Rev 22(1):57–77. doi:10.1002/mas.10042
58. Wiesner J, Premsler T, Sickmann A (2008) Application of electron transfer dissociation (ETD) for the analysis of posttranslational

modifications. Proteomics 8(21):4466–4483. doi:10.1002/pmic.200800329
59. An HJ, Froehlich JW, Lebrilla CB (2009) Determination of glycosylation sites and site-specific heterogeneity in glycoproteins. Curr Opin Chem Biol 13(4):421–426. doi:10.1016/j.cbpa.2009.07.022
60. Boersema PJ, Mohammed S, Heck AJR (2009) Phosphopeptide fragmentation and analysis by mass spectrometry. J Mass Spectrom 44(6):861–878. doi:10.1002/jms.1599
61. Schreiber TB, Mausbacher N, Breitkopf SB, Grundner-Culemann K, Daub H (2008) Quantitative phosphoproteomics: an emerging key technology in signal-transduction research. Proteomics 8(21):4416–4432. doi:10.1002/pmic.200800132
62. Crowe MC, Brodbelt JS (2004) Infrared multiphoton dissociation (IRMPD) and collisionally activated dissociation of peptides in a quadrupole ion trap with selective IRMPD of phosphopeptides. J Am Soc Mass Spectrom 15(11):1581–1592. doi:10.1016/j.jasms.2004.07.016
63. Crowe MC, Brodbelt JS (2005) Differentiation of phosphorylated and unphosphorylated peptides by high-performance liquid chromatography-electrospray ionization-infrared multiphoton dissociation in a quadrupole ion trap. Anal Chem 77(17):5726–5734. doi:10.1021/ac0509410
64. Brodbelt JS, Wilson JJ (2009) Infrared multiphoton dissociation in quadrupole ion traps. Mass Spectrom Rev 28(3):390–424. doi:10.1002/mas.20216
65. Little DP, Speir JP, Senko MW, Oconnor PB, McLafferty FW (1994) Infrared multiphoton dissociation of large multiply-charged ions for biomolecule sequencing. Anal Chem 66(18):2809–2815. doi:10.1021/ac00090a004
66. Ly T, Julian RR (2009) Ultraviolet photodissociation: developments towards applications for mass-spectrometry-based proteomics. Angew Chem Int Ed 48(39):7130–7137. doi:10.1002/anie.200900613
67. Reilly JP (2009) Ultraviolet photofragmentation of biomolecular ions. Mass Spectrom Rev 28(3):425–447. doi:10.1002/mas.20214
68. Gatlin CL, Eng JK, Cross ST, Detter JC, Yates JR (2000) Automated identification of amino acid sequence variations in proteins by HPLC/microspray tandem mass spectrometry. Anal Chem 72(4):757–763. doi:10.1021/ac991025n
69. Masselon C, Anderson GA, Harkewicz R, Bruce JE, Pasa-Tolic L, Smith RD (2000) Accurate mass multiplexed tandem mass spectrometry for high-throughput polypeptide identification from mixtures. Anal Chem 72(8):1918–1924. doi:10.1021/ac991133+
70. Purvine S, Eppel JT, Yi EC, Goodlett DR (2003) Shotgun collision-induced dissociation of peptides using a time of flight mass analyzer. Proteomics 3(6):847–850. doi:10.1002/pmic.200300362
71. Silva JC, Denny R, Dorschel CA, Gorenstein M, Kass IJ, Li GZ, McKenna T, Nold MJ, Richardson K, Young P, Geromanos S (2005) Quantitative proteomic analysis by accurate mass retention time pairs. Anal Chem 77(7):2187–2200. doi:10.1021/ac048455k
72. Geiger T, Cox J, Mann M (2010) Proteomics on an orbitrap benchtop mass spectrometer using All-ion fragmentation. Mol Cell Proteomics 9(10):2252–2261. doi:10.1074/mcp.M110.001537
73. Li LJ, Masselon CD, Anderson GA, Pasa-Tolic L, Lee SW, Shen YF, Zhao R, Lipton MS, Conrads TP, Tolic N, Smith RD (2001) High-throughput peptide identification from protein digests using data-dependent multiplexed tandem FTICR mass spectrometry coupled with capillary liquid chromatography. Anal Chem 73(14):3312–3322. doi:10.1021/ac010192w
74. Panchaud A, Scherl A, Shaffer SA, von Haller PD, Kulasekara HD, Miller SI, Goodlett DR (2009) Precursor acquisition independent from ion count: how to dive deeper into the proteomics ocean. Anal Chem 81(15):6481–6488. doi:10.1021/ac900888s
75. Venable JD, Dong MQ, Wohlschlegel J, Dillin A, Yates JR (2004) Automated approach for quantitative analysis of complex peptide mixtures from tandem mass spectra. Nat Methods 1(1):39–45. doi:10.1038/nmeth705
76. Panchaud A, Jung S, Shaffer SA, Aitchison JD, Goodlett DR (2011) Faster, quantitative, and accurate precursor acquisition independent from ion count. Anal Chem 83(6):2250–2257. doi:10.1021/ac103079q
77. Davis MT, Spahr CS, McGinley MD, Robinson JH, Bures EJ, Beierle J, Mort J, Yu W, Luethy R, Patterson SD (2001) Towards defining the urinary proteome using liquid chromatography-tandem mass spectrometry - II. Limitations of complex mixture analyses. Proteomics 1(1):108–117. doi:10.1002/1615-9861(200101)1:1<108::aid-prot108>3.0.co;2-5
78. Patterson SD, Spahr CS, Daugas E, Susin SA, Irinopoulou T, Koehler C, Kroemer G (2000) Mass spectrometric identification of proteins released from mitochondria undergoing permeability transition. Cell Death Differ 7(2):137–144. doi:10.1038/sj.cdd.4400640
79. Spahr CS, Davis MT, McGinley MD, Robinson JH, Bures EJ, Beierle J, Mort J, Courchesne PL, Chen K, Wahl RC, Yu W, Luethy R, Patterson SD (2001) Towards defining the urinary proteome using liquid chromatography-tandem mass spectrometry I. Profiling an unfractionated tryptic digest. Proteomics 1(1):93–107. doi:10.1002/1615-9861(200101)1:1<93::aid-prot93>3.0.co;2-3
80. Yi EC, Marelli M, Lee H, Purvine SO, Aebersold R, Aitchison JD, Goodlett DR (2002) Approaching complete peroxisome characterization by gas-phase fractionation. Electrophoresis 23(18):3205–3216. doi:10.1002/1522-2683(200209)23:18<3205::aid-elps3205>3.0.co;2-y
81. Scherl A, Shaffer SA, Taylor GK, Kulasekara HD, Miller SI, Goodlett DR (2008) Genome-specific gas-phase fractionation strategy for improved

shotgun proteomic profiling of proteotypic peptides. Anal Chem 80(4):1182–1191. doi:10.1021/ac701680f
82. Harvey SR, MacPhee CE, Barran PE (2011) Ion mobility mass spectrometry for peptide analysis. Methods 54(4):454–461. doi:10.1016/j.ymeth.2011.05.004
83. Valentine SJ, Kulchania M, Barnes CAS, Clemmer DE (2001) Multidimensional separations of complex peptide mixtures: a combined high-performance liquid chromatography/ion mobility/time-of-flight mass spectrometry approach. Int J Mass Spectrom 212(1–3):97–109. doi:10.1016/s1387-3806(01)00511-5
84. Srebalus CA, Li JW, Marshall WS, Clemmer DE (1999) Gas phase separations of electrosprayed peptide libraries. Anal Chem 71(18):3918–3927. doi:10.1021/ac9903757
85. Shvartsburg AA, Danielson WF, Smith RD (2010) High-resolution differential ion mobility separations using helium-rich gases. Anal Chem 82(6):2456–2462. doi:10.1021/ac902852a
86. Shvartsburg AA, Tang KQ, Smith RD (2010) Differential ion mobility separations of peptides with resolving power exceeding 50. Anal Chem 82(1):32–35. doi:10.1021/ac902133n
87. Shvartsburg AA, Prior DC, Tang KQ, Smith RD (2010) High-resolution differential ion mobility separations using planar analyzers at elevated dispersion fields. Anal Chem 82(18):7649–7655. doi:10.1021/ac101413k
88. Shvartsburg AA, Li FM, Tang KQ, Smith RD (2006) High-resolution field asymmetric waveform ion mobility spectrometry using new planar geometry analyzers. Anal Chem 78(11):3706–3714. doi:10.1021/ac052020v
89. Shvartsburg AA, Singer D, Smith RD, Hoffmann R (2011) Ion mobility separation of isomeric phosphopeptides from a protein with variant modification of adjacent residues. Anal Chem 83(13):5078–5085. doi:10.1021/ac200985s
90. Giles K, Pringle SD, Worthington KR, Little D, Wildgoose JL, Bateman RH (2004) Applications of a travelling wave-based radio-frequencyonly stacked ring ion guide. Rapid Commun Mass Spectrom 18(20):2401–2414. doi:10.1002/rcm.1641
91. Pringle SD, Giles K, Wildgoose JL, Williams JP, Slade SE, Thalassinos K, Bateman RH, Bowers MT, Scrivens JH (2007) An investigation of the mobility separation of some peptide and protein ions using a new hybrid quadrupole/travelling wave IMS/oa-ToF instrument. Int J Mass Spectrom 261(1):1–12. doi:10.1016/j.ijms.2006.07.021
92. Schmidt A, Claassen M, Aebersold R (2009) Directed mass spectrometry: towards hypothesis-driven proteomics. Curr Opin Chem Biol 13(5–6):510–517. doi:10.1016/j.cbpa.2009.08.016
93. Paulovich AG, Whiteaker JR, Hoofnagle AN, Wang P (2008) The interface between biomarker discovery and clinical validation: the tar pit of the protein biomarker pipeline. Proteomics Clin Appl 2(10–11):1386–1402. doi:10.1002/prca.200780174
94. Picotti P, Bodenmiller B, Mueller LN, Domon B, Aebersold R (2009) Full dynamic range proteome analysis of S. Cerevisiae by targeted proteomics. Cell 138(4):795–806. doi:10.1016/j.cell.2009.05.051
95. Fusaro VA, Mani DR, Mesirov JP, Carr SA (2009) Prediction of high-responding peptides for targeted protein assays by mass spectrometry. Nat Biotechnol 27(2):190–198. doi:10.1038/nbt.1524
96. Mallick P, Schirle M, Chen SS, Flory MR, Lee H, Martin D, Raught B, Schmitt R, Werner T, Kuster B, Aebersold R (2007) Computational prediction of proteotypic peptides for quantitative proteomics. Nat Biotechnol 25(1):125–131. doi:10.1038/nbt1275
97. Deutsch EW, Lam H, Aebersold R (2008) PeptideAtlas: a resource for target selection for emerging targeted proteomics workflows. EMBO Rep 9(5):429–434. doi:10.1038/embor.2008.56
98. Farrah T, Deutsch EW, Omenn GS, Campbell DS, Sun Z, Bletz JA, Mallick P, Katz JE, Malmstrom J, Ossola R, Watts JD, Lin BAY, Zhang H, Moritz RL, Aebersold R (2011) A high-confidence human plasma proteome reference set with estimated concentrations in PeptideAtlas. Mol Cell Proteomics 10(9). doi:10.1074/mcp.M110.006353
99. Kiyonami R, Schoen A, Prakash A, Peterman S, Zabrouskov V, Picotti P, Aebersold R, Huhmer A, Domon B (2011) Increased selectivity, analytical precision, and throughput in targeted proteomics. Mol Cell Proteomics 10(2). doi:10.1074/mcp.M110.002931
100. Addona TA, Abbatiello SE, Schilling B, Skates SJ, Mani DR, Bunk DM, Spiegelman CH, Zimmerman LJ, Ham AJ, Keshishian H, Hall SC, Allen S, Blackman RK, Borchers CH, Buck C, Cardasis HL, Cusack MP, Dodder NG, Gibson BW, Held JM, Hiltke T, Jackson A, Johansen EB, Kinsinger CR, Li J, Mesri M, Neubert TA, Niles RK, Pulsipher TC, Ransohoff D, Rodriguez H, Rudnick PA, Smith D, Tabb DL, Tegeler TJ, Variyath AM, Vega-Montoto LJ, Wahlander A, Waldemarson S, Wang M, Whiteaker JR, Zhao L, Anderson NL, Fisher SJ, Liebler DC, Paulovich AG, Regnier FE, Tempst P, Carr SA (2009) Multi-site assessment of the precision and reproducibility of multiple reaction monitoring-based measurements of proteins in plasma. Nat Biotechnol 27(7):633–641. doi:nbt.1546 [pii] 10.1038/nbt.1546
101. Calvo E, Camafeita E, Fernandez-Gutierrez B, Lopez JA (2011) Applying selected reaction monitoring to targeted proteomics. Expert Rev Proteomics 8(2):165–173. doi:10.1586/epr.11.11
102. Chiu CL, Randall S, Molloy MP (2009) Recent progress in selected reaction monitoring MS-driven plasma protein biomarker analysis. Bioanalysis 1(4):847–855. doi:10.4155/bio.09.56

103. Elschenbroich S, Kislinger T (2011) Targeted proteomics by selected reaction monitoring mass spectrometry: applications to systems biology and biomarker discovery. Mol Biosyst 7(2):292–303. doi:10.1039/c0mb00159g
104. Surinova S, Schiess R, Huttenhain R, Cerciello F, Wollscheid B, Aebersold R (2011) On the development of plasma protein biomarkers. J Proteome Res 10(1):5–16. doi:10.1021/pr1008515
105. Martin DB, Holzman T, May D, Peterson A, Eastham A, Eng J, McIntosh M (2008) MRMer, an interactive open source and cross-platform system for data extraction and visualization of multiple reaction monitoring experiments. Mol Cell Proteomics 7(11):2270–2278. doi:10.1074/mcp.M700504-MCP200
106. Mead JA, Bianco L, Ottone V, Barton C, Kay RG, Lilley KS, Bond NJ, Bessant C (2009) MRMaid, the web-based tool for designing multiple reaction monitoring (MRM) transitions. Mol Cell Proteomics 8(4):696–705. doi:10.1074/mcp.M800192-MCP200
107. Sherwood CA, Eastham A, Lee LW, Peterson A, Eng JK, Shteynberg D, Mendoza L, Deutsch EW, Risler J, Tasman N, Aebersold R, Lam H, Martin DB (2009) MaRiMba: a software application for spectral library-based MRM transition list assembly. J Proteome Res 8(10):4396–4405. doi:10.1021/pr900010h
108. MacLean B, Tomazela DM, Shulman N, Chambers M, Finney GL, Frewen B, Kern R, Tabb DL, Liebler DC, MacCoss MJ (2010) Skyline: an open source document editor for creating and analyzing targeted proteomics experiments. Bioinformatics 26(7):966–968. doi:10.1093/bioinformatics/btq054
109. MacLean B, Tomazela DM, Abbatiello SE, Zhang SC, Whiteaker JR, Paulovich AG, Carr SA, MacCoss MJ (2010) Effect of collision energy optimization on the measurement of peptides by selected reaction monitoring (SRM) mass spectrometry. Anal Chem 82(24):10116–10124. doi:10.1021/ac102179j
110. Prakash A, Tomazela DM, Frewen B, MacLean B, Merrihew G, Peterman S, MacCoss MJ (2009) Expediting the development of targeted SRM assays: using data from shotgun proteomics to automate method development. J Proteome Res 8(6):2733–2739. doi:10.1021/pr801028b
111. Chait BT, Wang R, Beavis RC, Kent SBH (1993) Protein ladder sequencing. Science 262(5130):89–92
112. Cruz-Marcelo A, Guerra R, Vannucci M, Li Y, Lau CC, Man TK (2008) Comparison of algorithms for pre-processing of SELDI-TOF mass spectrometry data. Bioinformatics 24(19):2129–2136. doi:btn398 [pii] 10.1093/bioinformatics/btn398
113. Roy P, Truntzer C, Maucort-Boulch D, Jouve T, Molinari N (2011) Protein mass spectra data analysis for clinical biomarker discovery: a global review. Brief Bioinform 12(2):176–186. doi:bbq019 [pii] 10.1093/bib/bbq019
114. Sellers KF, Miecznikowski JC (2010) Feature detection techniques for preprocessing proteomic data. Int J Biomed Imaging 2010:896718. doi:10.1155/2010/896718
115. Wegdam W, Moerland PD, Buist MR, Loren V, van Themaat E, Bleijlevens B, Hoefsloot HC, de Koster CG, Aerts JM (2009) Classification-based comparison of pre-processing methods for interpretation of mass spectrometry generated clinical datasets. Proteome Sci 7:19. doi:1477-5956-7-19 [pii] 10.1186/1477-5956-7-19
116. Addona T, Clauser K (2002) De Novo Peptide De Novo Peptide Sequencing via Manual Interpretation of MS/MS Spectra. Curr Protoc Protein Sci 16.11.1–16.11.19
117. Craig R, Beavis RC (2004) TANDEM: matching proteins with tandem mass spectra. Bioinformatics 20(9):1466–1467. doi:10.1093/bioinformatics/bth092
118. Geer LY, Markey SP, Kowalak JA, Wagner L, Xu M, Maynard DM, Yang X, Shi W, Bryant SH (2004) Open mass spectrometry search algorithm. J Proteome Res 3(5):958–964. doi:10.1021/pr0499491
119. Cox J, Neuhauser N, Michalski A, Scheltema RA, Olsen JV, Mann M (2011) Andromeda: a peptide search engine integrated into the MaxQuant environment. J Proteome Res 10(4):1794–1805. doi:10.1021/Pr101065j
120. Tabb DL, Fernando CG, Chambers MC (2007) MyriMatch: highly accurate tandem mass spectral peptide identification by multivariate hypergeometric analysis. J Proteome Res 6(2):654–661. doi:10.1021/Pr0604054
121. Clauser KR, Baker P, Burlingame AL (1999) Role of accurate mass measurement (+/− 10 ppm) in protein identification strategies employing MS or MS MS and database searching. Anal Chem 71(14):2871–2882
122. Colinge J, Masselot A, Giron M, Dessingy T, Magnin J (2003) OLAV: towards high-throughput tandem mass spectrometry data identification. Proteomics 3(8):1454–1463. doi:10.1002/pmic.200300485
123. Casado-Vela J (2011) Lights and shadows of proteomic technologies for the study of protein species including isoforms, splicing variants and protein post-translational modifications (vol 11, pg 590, 2011). Proteomics 11(7):1370–1370
124. Nilsson T, Mann M, Aebersold R, Yates JR, Bairoch A, Bergeron JJM (2010) Mass spectrometry in high-throughput proteomics: ready for the big time. Nat Methods 7(9):681–685. doi:10.1038/nmeth0910-681
125. Searle BC, Turner M, Nesvizhskii AI (2008) Improving sensitivity by probabilistically combining results from multiple MS/MS search methodologies. J Proteome Res 7(1):245–253. doi:10.1021/Pr070540w
126. Resing KA, Meyer-Arendt K, Mendoza AM, Aveline-Wolf LD, Jonscher KR, Pierce KG, Old

WM, Cheung HT, Russell S, Wattawa JL, Goehle GR, Knight RD, Ahn NG (2004) Improving reproducibility and sensitivity in identifying human proteins by shotgun proteomics. Anal Chem 76(13):3556–3568
127. Alves G, Wu WW, Wang GH, Shen RF, Yu YK (2008) Enhancing peptide identification confidence by combining search methods. J Proteome Res 7(8):3102–3113. doi:10.1021/Pr700798h
128. Searle BC, Turner M (2006) Improving computer interpretation of linear ion trap proteomics data using Scaffold. Mol Cell Proteomics 5(10): S297–S297
129. Searle BC (2010) Scaffold: a bioinformatic tool for validating MS/MS-based proteomic studies. Proteomics 10(6):1265–1269. doi:10.1002/pmic.200900437
130. Kwon T, Choi H, Vogel C, Nesvizhskii AI, Marcotte EM (2011) MSblender: a probabilistic approach for integrating peptide identifications from multiple database search engines. J Proteome Res 10(7):2949–2958. doi:10.1021/Pr2002116
131. Yates JR, Morgan SF, Gatlin CL, Griffin PR, Eng JK (1998) Method to compare collision-induced dissociation spectra of peptides: potential for library searching and subtractive analysis. Anal Chem 70(17):3557–3565
132. Lam H, Deutsch EW, Eddes JS, Eng JK, King N, Stein SE, Aebersold R (2007) Development and validation of a spectral library searching method for peptide identification from MS/MS. Proteomics 7(5):655–667
133. Lam H, Deutsch EW, Eddes JS, Eng JK, Stein SE, Aebersold R (2008) Building consensus spectral libraries for peptide identification in proteomics. Nat Methods 5(10):873–875. doi:10.1038/Nmeth.1254
134. Frewen BE, Merrihew GE, Wu CC, Noble WS, MacCoss MJ (2006) Analysis of peptide MS/MS spectra from large-scale proteomics experiments using spectrum libraries. Anal Chem 78(16):5678–5684. doi:10.1021/Ac060279n
135. Craig R, Cortens JC, Fenyo D, Beavis RC (2006) Using annotated peptide mass spectrum libraries for protein identification. J Proteome Res 5(8):1843–1849. doi:10.1021/Pr0602085
136. Hummel J, Niemann M, Wienkoop S, Schulze W, Steinhauser D, Selbig J, Walther D, Weckwerth W (2007) ProMEX: a mass spectral reference database for proteins and protein phosphorylation sites. BMC Bioinformatics 8. doi:Artn 216 Doi 10.1186/1471-2105-8-216
137. Wu X, Tseng CW, Edwards N (2007) HMMatch: peptide identification by spectral matching of tandem mass spectra using hidden Markov models. J Comput Biol 14(8):1025–1043. doi:10.1089/cmb.2007.0071
138. Bodenmiller B, Campbell D, Gerrits B, Lam H, Jovanovic M, Picotti P, Schlapbach R, Aebersold R (2008) PhosphoPep-a database of protein phosphorylation sites in model organisms. Nat Biotechnol 26(12):1339–1340. doi:10.1038/Nbt1208-1339
139. Srikumar T, Jeram SM, Lam H, Raught B (2010) A ubiquitin and ubiquitin-like protein spectral library. Proteomics 10(2):337–342. doi:10.1002/pmic.200900627
140. Desiere F, Deutsch EW, King NL, Nesvizhskii AI, Mallick P, Eng J, Chen S, Eddes J, Loevenich SN, Aebersold R (2006) The PeptideAtlas project. Nucleic Acids Res 34(Database issue):D655–D658
141. Vizcaino JA, Cote R, Reisinger F, Foster JM, Mueller M, Rameseder J, Hermjakob H, Martens L (2009) A guide to the proteomics identifications database proteomics data repository. Proteomics 9(18):4276–4283. doi:10.1002/pmic.200900402
142. Jones P, Cote RG, Martens L, Quinn AF, Taylor CF, Derache W, Hermjakob H, Apweiler R (2006) PRIDE: a public repository of protein and peptide identifications for the proteomics community. Nucleic Acids Res 34:D659–D663. doi:10.1093/Nar/Gkj138
143. Nesvizhskii AI (2010) A survey of computational methods and error rate estimation procedures for peptide and protein identification in shotgun proteomics. J Proteomics 73(11):2092–2123. doi:S1874-3919(10)00249-6 [pii] 10.1016/j.jprot.2010.08.009
144. Frank AM (2009) Predicting intensity ranks of peptide fragment ions. J Proteome Res 8(5):2226–2240. doi:10.1021/Pr800677f
145. Cox J, Mann M (2009) Computational principles of determining and improving mass precision and accuracy for proteome measurements in an orbitrap. J Am Soc Mass Spectrom 20(8):1477–1485. doi:10.1016/j.jasms.2009.05.007
146. Hamm CW, Wilson WE, Harvan DJ (1986) Peptide sequencing program. Comput Appl Biosci 2(2):115–118
147. Ishikawa K, Niwa Y (1986) Computer-aided peptide sequencing by fast-atom-bombardment mass-spectrometry. Biomed Environ Mass 13(7): 373–380
148. Sakurai T, Matsuo T, Matsuda H, Katakuse I (1984) Paas-3: a computer-program to determine probable sequence of peptides from mass-spectrometric data. Biomed Mass Spectrom 11(8):396–399
149. Scoble HA, Biller JE, Biemann K (1987) A graphics display-oriented strategy for the amino-acid sequencing of peptides by tandem mass-spectrometry. Fresen Z Anal Chem 327(2):239–245
150. Siegel MM, Bauman N (1988) An efficient algorithm for sequencing peptides using fast atom bombardment mass-spectral data. Biomed Environ Mass 15(6):333–343
151. Bartels C (1990) Fast algorithm for peptide sequencing by mass-spectroscopy. Biomed Environ Mass 19(6):363–368
152. Taylor JA, Johnson RS (2001) Implementation and uses of automated de novo peptide sequencing by tandem mass spectrometry. Anal Chem 73(11):2594–2604

153. Dancik V, Addona TA, Clauser KR, Vath JE, Pevzner PA (1999) De novo peptide sequencing via tandem mass spectrometry. J Comput Biol 6(3–4):327–342
154. Chen T, Kao MY, Tepel M, Rush J, Church GM (2001) A dynamic programming approach to de novo peptide sequencing via tandem mass spectrometry. J Comput Biol 8(3):325–337
155. Frank A, Pevzner P (2005) PepNovo: De novo peptide sequencing via probabilistic network modeling. Anal Chem 77(4):964–973. doi:10.1021/Ac048788h
156. Fischer B, Roth V, Roos F, Grossmann J, Baginsky S, Widmayer P, Gruissem W, Buhmann JM (2005) NovoHMM: a hidden Markov model for de novo peptide sequencing. Anal Chem 77(22):7265–7273. doi:10.1021/Ac0508853
157. Tabb DL, Ma ZQ, Martin DB, Ham AJL, Chambers MC (2008) DirecTag: accurate sequence tags from peptide MS/MS through statistical scoring. J Proteome Res 7(9):3838–3846. doi:10.1021/Pr800154p
158. Ma B, Zhang K, Hendrie C, Liang C, Li M, Kirby AD, Lajoie G (2003) Peaks: powerful software for peptide de novo sequencing by tandem mass spectrometry. Rapid Commun Mass Spectrom 17:2337–2342
159. Chi H, Sun RX, Yang B, Song CQ, Wang LH, Liu C, Fu Y, Yuan ZF, Wang HP, He SM, Dong MQ (2010) PNovo: De novo peptide sequencing and identification using HCD spectra. J Proteome Res 9(5):2713–2724. doi:10.1021/Pr100182k
160. Zhang ZQ, McElvain JS (2000) De novo peptide sequencing by two dimensional fragment correlation mass spectrometry. Anal Chem 72(11):2337–2350
161. Savitski MM, Nielsen ML, Kjeldsen F, Zubarev RA (2005) Proteomics-grade de novo sequencing approach. J Proteome Res 4(6):2348–2354. doi:10.1021/pr050288x
162. Bandeira N, Tsur D, Frank A, Pevzner PA (2007) Protein identification by spectral networks analysis. Proc Natl Acad Sci U S A 104(15):6140–6145. doi:10.1073/pnas.0701130104
163. Datta R, Bern M (2009) Spectrum fusion: using multiple mass spectra for de novo peptide sequencing. J Comput Biol 16(8):1169–1182. doi:10.1089/cmb.2009.0122
164. Mann M, Wilm M (1994) Error tolerant identification of peptides in sequence databases by peptide sequence tags. Anal Chem 66(24):4390–4399
165. Tabb DL, Saraf A, Yates JR (2003) GutenTag: high-throughput sequence tagging via an empirically derived fragmentation model. Anal Chem 75(23):6415–6421. doi:10.1021/Ac0347462
166. Tanner S, Shu HJ, Frank A, Wang LC, Zandi E, Mumby M, Pevzner PA, Bafna V (2005) InsPecT: identification of posttranslationally modified peptides from tandem mass spectra. Anal Chem 77(14):4626–4639. doi:10.1021/Ac050102d
167. Dasari S, Chambers MC, Slebos RJ, Zimmerman LJ, Ham AJL, Tabb DL (2010) TagRecon: high-throughput mutation identification through sequence tagging. J Proteome Res 9(4):1716–1726. doi:10.1021/pr900850m
168. Shilov IV, Seymour SL, Patel AA, Loboda A, Tang WH, Keating SP, Hunter CL, Nuwaysir LM, Schaeffer DA (2007) The paragon algorithm, a next generation search engine that uses sequence temperature values and feature probabilities to identify peptides from tandem mass spectra. Mol Cell Proteomics 6(9):1638–1655. doi:10.1074/mcp.T600050-MCP200
169. Nesvizhskii AI, Aebersold R (2005) Interpretation of shotgun proteomic data: the protein inference problem. Mol Cell Proteomics 4(10):1419–1440
170. Nesvizhskii AI, Aebersold R (2004) Analysis, statistical validation and dissemination of large-scale proteomics datasets generated by tandem MS. Drug Discov Today 9(4):173–181. doi:10.1016/S1359-6446(03)02978-7 S1359644603029787 [pii]
171. States DJ, Omenn GS, Blackwell TW, Fermin D, Eng J, Speicher DW, Hanash SM (2006) Challenges in deriving high-confidence protein identifications from data gathered by a HUPO plasma proteome collaborative study. Nat Biotechnol 24(3):333–338. doi:10.1038/Nbt1183
172. Benjamini Y, Hochberg Y (1995) Controlling the false discovery rate–a practical and powerful approach to multiple testing. J Roy Stat Soc B Met 57(1):289–300
173. Keller A, Nesvizhskii AI, Kolker E, Aebersold R (2002) Empirical statistical model to estimate the accuracy of peptide identifications made by ms/ms and database search. Anal Chem 74(20):5383–5392
174. Bianco L, Mead JA, Bessant C (2009) Comparison of novel decoy database designs for optimizing protein identification searches using ABRF sPRG2006 standard MS/MS data sets. J Proteome Res 8(4):1782–1791. doi:10.1021/Pr800792z
175. Kall L, Storey JD, MacCoss MJ, Noble WS (2008) Assigning significance to peptides identified by tandem mass spectrometry using decoy databases. J Proteome Res 7(1):29–34. doi:10.1021/pr700600n
176. Storey JD, Tibshirani R (2003) Statistical significance for genomewide studies. Proc Natl Acad Sci U S A 100(16):9440–9445. doi:10.1073/pnas.1530509100 1530509100 [pii]
177. Qian W-J, Liu T, Monroe ME, Strittmatter EF, Jacobs JM, Kangas LJ, Petritis K, Camp DG, Smith RD (2005) Probability-based evaluation of peptide and protein identifications from tandem mass spectrometry and SEQUEST analysis: the human proteome. J Proteome Res 4(1):53–62
178. Moore RE, Young MK, Lee TD (2002) Qscore: an algorithm for evaluating SEQUEST database search results. J Am Soc Mass Spectrom 13(4):378–386
179. Nesvizhskii AI, Keller A, Kolker E, Aebersold R (2003) A statistical model for identifying proteins by tandem mass spectrometry. Anal Chem 75(17):4646–4658

180. Ma ZQ, Dasari S, Chambers MC, Litton MD, Sobecki SM, Zimmerman LJ, Halvey PJ, Schilling B, Drake PM, Gibson BW, Tabb DL (2009) IDPicker 2.0: improved protein assembly with high discrimination peptide identification filtering. J Proteome Res 8(8):3872–3881. doi:10.1021/pr900360j
181. Kim S, Mischerikow N, Bandeira N, Navarro JD, Wich L, Mohammed S, Heck AJ, Pevzner PA (2010) The generating function of CID, ETD and CID/ETD pairs of tandem mass spectra: applications to database search. Mol Cell Proteomics. doi:M110.003731 [pii] 10.1074/mcp.M110.003731
182. Fenyo D, Beavis RC (2003) A method for assessing the statistical significance of mass spectrometry-based protein identifications using general scoring schemes. Anal Chem 75(4):768–774. doi:10.1021/Ac0258709
183. Taylor CF (2006) A capital workshop for the HUPO proteomics standards initiative. J Proteome Res 5(12):3229–3230
184. Hermjakob H (2006) The HUPO proteomics standards initiative–overcoming the fragmentation of proteomics data. Proteomics 6(1):34–38. doi:10.1002/pmic.200600537
185. Taylor CF, Hermjakob H, Julian RK, Garavelli JS, Aebersold R, Apweiler R (2006) The work of the human proteome organisation's proteomics standards initiative (HUPO PSI). Omics 10(2):145–151
186. Orchard S, Kersey P, Hermjakob H, Apweiler R (2003) Meeting review: The HUPO proteomics standards initiative meeting: towards common standards for exchanging proteomics data—Hinxton, Cambridge, UK, 19–20 October 2002. Comp Funct Genom 4(1):16–19. doi:10.1002/Cfg.232
187. Orchard S, Kersey P, Zhu WM, Montecchi-Palazzi L, Hermjakob H, Apweiler R (2003) Meeting review: progress in establishing common standards for exchanging proteomics data: the second meeting of the HUPO proteomics standards initiative. Comp Funct Genom 4(2):203–206. doi:10.1002/Cfg.279
188. Bradshaw RA, Burlingame AL, Carr S, Aebersold R (2006) Reporting protein identification data–the next generation of guidelines. Mol Cell Proteomics 5(5):787–788
189. Becker GW (2008) Stable isotopic labeling of proteins for quantitative proteomic applications. Brief Funct Genomic Proteomic 7(5):371–382. doi:eln047 [pii] 10.1093/bfgp/eln047
190. Gevaert K, Impens F, Ghesquiere B, Van Damme P, Lambrechts A, Vandekerckhove J (2008) Stable isotopic labeling in proteomics. Proteomics 8(23–24):4873–4885. doi:10.1002/pmic.200800421
191. Oda Y, Huang K, Cross FR, Cowburn D, Chait BT (1999) Accurate quantitation of protein expression and site-specific phosphorylation. Proc Natl Acad Sci U S A 96(12):6591–6596
192. Ong SE, Blagoev B, Kratchmarova I, Kristensen DB, Steen H, Pandey A, Mann M (2002) Stable isotope labeling by amino acids in cell culture, SILAC, as a simple and accurate approach to expression proteomics. Mol Cell Proteomics 1(5):376–386
193. Gygi SP, Rist B, Gerber SA, Turecek F, Gelb MH, Aebersold R (1999) Quantitative analysis of complex protein mixtures using isotope-coded affinity tags. Nat Biotechnol 17(10):994–999
194. Dayon L, Hainard A, Licker V, Turck N, Kuhn K, Hochstrasser DF, Burkhard PR, Sanchez JC (2008) Relative quantification of proteins in human cerebrospinal fluids by MS/MS using 6-plex isobaric tags. Anal Chem 80(8):2921–2931. doi:10.1021/ac702422x
195. Ross PL, Huang YN, Marchese JN, Williamson B, Parker K, Hattan S, Khainovski N, Pillai S, Dey S, Daniels S, Purkayastha S, Juhasz P, Martin S, Bartlet-Jones M, He F, Jacobson A, Pappin DJ (2004) Multiplexed protein quantitation in *Saccharomyces cerevisiae* using amine-reactive isobaric tagging reagents. Mol Cell Proteomics 3(12): 1154–1169
196. Reynolds KJ, Fenselau C (2004) Quantitative protein analysis using proteolytic [18O] water labeling. Curr Protoc Protein Sci 23:23–24. doi:10.1002/0471140864.ps2304s34
197. Ow SY, Cardona T, Taton A, Magnuson A, Lindblad P, Stensjo K, Wright PC (2008) Quantitative shotgun proteomics of enriched heterocysts from Nostoc sp. PCC 7120 using 8-plex isobaric peptide tags. J Proteome Res 7(4):1615–1628. doi:10.1021/pr700604v
198. Wu CC, MacCoss MJ, Howell KE, Matthews DE, Yates JR 3rd (2004) Metabolic labeling of mammalian organisms with stable isotopes for quantitative proteomic analysis. Anal Chem 76(17): 4951–4959
199. Savitski MM, Sweetman G, Askenazi M, Marto JA, Lang M, Zinn N, Bantscheff M (2011) Delayed fragmentation and optimized isolation width settings for improvement of protein identification and accuracy of isobaric mass tag quantification on orbitrap-type mass spectrometers. Anal Chem 83(23):8959–8967. doi:10.1021/ac201760x
200. Neilson KA, Ali NA, Muralidharan S, Mirzaei M, Mariani M, Assadourian G, Lee A, van Sluyter SC, Haynes PA (2011) Less label, more free: approaches in label-free quantitative mass spectrometry. Proteomics 11(4):535–553. doi:10.1002/pmic.201000553
201. Sardiu ME, Cai Y, Jin J, Swanson SK, Conaway RC, Conaway JW, Florens L, Washburn MP (2008) Probabilistic assembly of human protein interaction networks from label-free quantitative proteomics. Proc Natl Acad Sci U S A 105(5):1454–1459. doi:0706983105 [pii] 10.1073/pnas.0706983105
202. Ishihama Y, Oda Y, Tabata T, Sato T, Nagasu T, Rappsilber J, Mann M (2005) Exponentially modified protein abundance index (emPAI) for estimation of absolute protein amount in proteomics by the number of sequenced peptides per protein. Mol

Cell Proteomics 4(9):1265–1272. doi:M500061-MCP200 [pii] 10.1074/mcp.M500061-MCP200
203. Rappsilber J, Ryder U, Lamond AI, Mann M (2002) Large-scale proteomic analysis of the human spliceosome. Genome Res 12(8):1231–1245. doi:10.1101/gr.473902
204. Mallick P, Schirle M, Chen SS, Flory MR, Lee H, Martin D, Ranish J, Raught B, Schmitt R, Werner T, Kuster B, Aebersold R (2007) Computational prediction of proteotypic peptides for quantitative proteomics. Nat Biotechnol 25(1):125–131. doi:nbt1275 [pii] 10.1038/nbt1275
205. Braisted JC, Kuntumalla S, Vogel C, Marcotte EM, Rodrigues AR, Wang R, Huang ST, Ferlanti ES, Saeed AI, Fleischmann RD, Peterson SN, Pieper R (2008) The APEX quantitative proteomics tool: generating protein quantitation estimates from LC-MS/MS proteomics results. BMC Bioinformatics 9:529. doi:1471-2105-9-529 [pii] 10.1186/1471-2105-9-529
206. Lundgren DH, Hwang SI, Wu L, Han DK (2010) Role of spectral counting in quantitative proteomics. Expert Rev Proteomics 7(1):39–53. doi:10.1586/epr.09.69
207. Paoletti AC, Parmely TJ, Tomomori-Sato C, Sato S, Zhu D, Conaway RC, Conaway JW, Florens L, Washburn MP (2006) Quantitative proteomic analysis of distinct mammalian Mediator complexes using normalized spectral abundance factors. Proc Natl Acad Sci U S A 103(50):18928–18933. doi:0606379103 [pii] 10.1073/pnas.0606379103
208. Mann K, Mann M (2008) The chicken egg yolk plasma and granule proteomes. Proteomics 8(1):178–191. doi:10.1002/pmic.200700790
209. Olson DG, Tripathi SA, Giannone RJ, Lo J, Caiazza NC, Hogsett DA, Hettich RL, Guss AM, Dubrovsky G, Lynd LR (2010) Deletion of the Cel48S cellulase from *Clostridium thermocellum*. Proc Natl Acad Sci U S A 107(41):17727–17732. doi:1003584107 [pii] 10.1073/pnas.1003584107
210. Park SK, Venable JD, Xu T, Yates JR 3rd (2008) A quantitative analysis software tool for mass spectrometry-based proteomics. Nat Methods 5(4):319–322. doi:nmeth.1195 [pii] 10.1038/nmeth.1195
211. Heinecke NL, Pratt BS, Vaisar T, Becker L (2010) PepC: proteomics software for identifying differentially expressed proteins based on spectral counting. Bioinformatics 26(12):1574–1575. doi:btq171 [pii] 10.1093/bioinformatics/btq171
212. Griffin NM, Yu J, Long F, Oh P, Shore S, Li Y, Koziol JA, Schnitzer JE (2010) Label-free, normalized quantification of complex mass spectrometry data for proteomic analysis. Nat Biotechnol 28(1):83–89. doi:nbt.1592 [pii] 10.1038/nbt.1592
213. Tsou CC, Tsai CF, Tsui YH, Sudhir PR, Wang YT, Chen YJ, Chen JY, Sung TY, Hsu WL (2010) IDEAL-Q, an automated tool for label-free quantitation analysis using an efficient peptide alignment approach and spectral data validation. Mol Cell Proteomics 9(1):131–144. doi:M900177-MCP200 [pii] 10.1074/mcp.M900177-MCP200
214. Cox J, Mann M (2008) MaxQuant enables high peptide identification rates, individualized p.p.b.-range mass accuracies and proteome-wide protein quantification. Nat Biotechnol 26(12):1367–1372. doi:nbt.1511 [pii] 10.1038/nbt.1511
215. Bellew M, Coram M, Fitzgibbon M, Igra M, Randolph T, Wang P, May D, Eng J, Fang R, Lin C, Chen J, Goodlett D, Whiteaker J, Paulovich A, McIntosh M (2006) A suite of algorithms for the comprehensive analysis of complex protein mixtures using high-resolution LC-MS. Bioinformatics 22(15):1902–1909. doi:btl276 [pii] 10.1093/bioinformatics/btl276
216. Pluskal T, Castillo S, Villar-Briones A, Oresic M (2010) MZmine 2: modular framework for processing, visualizing, and analyzing mass spectrometry-based molecular profile data. BMC Bioinformatics 11:395. doi:1471-2105-11-395 [pii] 10.1186/1471-2105-11-395
217. Jaffe JD, Mani DR, Leptos KC, Church GM, Gillette MA, Carr SA (2006) PEPPeR, a platform for experimental proteomic pattern recognition. Mol Cell Proteomics 5(10):1927–1941. doi:M600222-MCP200 [pii] 10.1074/mcp.M600222-MCP200
218. Mueller LN, Rinner O, Schmidt A, Letarte S, Bodenmiller B, Brusniak MY, Vitek O, Aebersold R, Muller M (2007) SuperHirn–a novel tool for high resolution LC-MS-based peptide/protein profiling. Proteomics 7(19):3470–3480. doi:10.1002/pmic.200700057
219. Bondarenko PV, Chelius D, Shaler TA (2002) Identification and relative quantitation of protein mixtures by enzymatic digestion followed by capillary reversed-phase liquid chromatography-tandem mass spectrometry. Anal Chem 74(18):4741–4749
220. Chelius D, Bondarenko PV (2002) Quantitative profiling of proteins in complex mixtures using liquid chromatography and mass spectrometry. J Proteome Res 1(4):317–323
221. Christin C, Bischoff R, Horvatovich P (2011) Data processing pipelines for comprehensive profiling of proteomics samples by label-free LC-MS for biomarker discovery. Talanta 83(4):1209–1224. doi:S0039-9140(10)00825-8 [pii] 10.1016/j.talanta.2010.10.029
222. Contrepois K, Ezan E, Mann C, Fenaille F (2010) Ultra-high performance liquid chromatography-mass spectrometry for the fast profiling of histone post-translational modifications. J Proteome Res 9(10):5501–5509. doi:10.1021/pr100497a
223. Callister SJ, Barry RC, Adkins JN, Johnson ET, Qian WJ, Webb-Robertson BJ, Smith RD, Lipton MS (2006) Normalization approaches for removing systematic biases associated with mass spectrometry and label-free proteomics. J Proteome Res 5(2):277–286. doi:10.1021/pr0503001
224. Kultima K, Nilsson A, Scholz B, Rossbach UL, Falth M, Andren PE (2009) Development and eval-

uation of normalization methods for label-free relative quantification of endogenous peptides. Mol Cell Proteomics 8(10):2285–2295. doi:M800514-MCP200 [pii] 10.1074/mcp. M800514-MCP200
225. Rajcevic U, Niclou SP, Jimenez CR (2009) Proteomics strategies for target identification and biomarker discovery in cancer. Front Biosci 14:3292–3303. doi:3452 [pii]
226. Zhu W, Smith JW, Huang CM (2010) Mass spectrometry-based label-free quantitative proteomics. J Biomed Biotechnol 2010:840518. doi:10.1155/2010/840518
227. Lengqvist J, Andrade J, Yang Y, Alvelius G, Lewensohn R, Lehtio J (2009) Robustness and accuracy of high speed LC-MS separations for global peptide quantitation and biomarker discovery. J Chromatogr B Analyt Technol Biomed Life Sci 877(13):1306–1316. doi:S1570-0232(09)00138-X [pii] 10.1016/j.jchromb.2009.02.052
228. Huang SK, Darfler MM, Nicholl MB, You J, Bemis KG, Tegeler TJ, Wang M, Wery JP, Chong KK, Nguyen L, Scolyer RA, Hoon DS (2009) LC/MS-based quantitative proteomic analysis of paraffin-embedded archival melanomas reveals potential proteomic biomarkers associated with metastasis. PLoS One 4(2):e4430. doi:10.1371/journal.pone.0004430
229. Huang JT, McKenna T, Hughes C, Leweke FM, Schwarz E, Bahn S (2007) CSF biomarker discovery using label-free nano-LC-MS based proteomic profiling: technical aspects. J Sep Sci 30(2):214–225
230. Pavelka N, Fournier ML, Swanson SK, Pelizzola M, Ricciardi-Castagnoli P, Florens L, Washburn MP (2008) Statistical similarities between transcriptomics and quantitative shotgun proteomics data. Mol Cell Proteomics 7(4):631–644. doi:M700240-MCP200 [pii] 10.1074/mcp.M700240-MCP200

Ionizing Radiation Effects on Cells, Organelles and Tissues on Proteome Level

2

Soile Tapio

Abstract

This chapter will review the proteome alterations induced by ionizing radiation in cellular systems or using animal models with whole body or localised exposure. The recent developments in qualitative and quantitative proteome analysis using formalin-fixed paraffin-embedded material from radiobiology archives will be illustrated. The development of promising protein targets to be used as radiation biomarkers in future molecular epidemiology studies is described.

Keywords

Proteome • Ionizing radiation • Endothelial cells • EA.hy926 • Lymphocytes • Nucleolar proteome • Cardiovascular system • Heart • Mitochondria • Formalin-fixed • Paraffin-embedded • Biomarker

2.1 Radiation Effects on the Endothelial Proteome

The vascular endothelium forms a continuous cellular monolayer lining the interior surface of blood vessels, and serves as the barrier between circulating blood and the subendothelial matrix [1]. It plays an important role in the integration and modulation of many functions of the arterial wall [2, 3]. Previous data show that increased production of reactive oxygen species (ROS) in the endothelium is associated with vascular endothelial dysfunction during the human aging process [4]. Increased ROS production leads to oxidation of low density lipoproteins (LDL), accumulation of lipids into foam cells, growth of vascular wall intima layer and finally atherosclerotic plaque expansion and rupture [5, 6].

Pluder et al. studied the immediate effects of low-dose (200 mGy) ionizing radiation (Co-60 gamma) using a human endothelial cell line EA.hy926 as the vascular model [7]. This cell line has proven to be a good endothelial model in other proteomic studies [8, 9] as it retains a primary endothelial cell-like transcriptome that also responds typically to statins that are widely used to reduce the risk of cardiovascular disease [10]. Moreover, as low radiation doses do not trigger

S. Tapio (✉)
Helmholtz Zentrum München, German Research Center for Environmental Health, Institute of Radiation Biology, Ingolstädter Landstrasse 1, Neuherberg 85764, Germany
e-mail: soile.tapio@helmholtz-muenchen.de

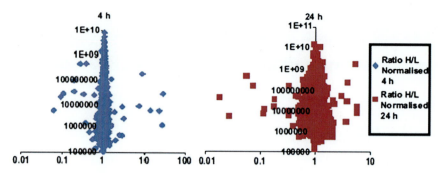

Fig. 2.1 "Christmas tree" model of all quantified proteins by SILAC. This figure shows normalized protein ratios plotted against summed peptide intensities. The proteins represented by data points lying close to the y-axis (y-axis = 1) did not show any expression changes. Outliers were considered as proteins with significantly differential expression only if they had a p-value < 0.01 and variability less than 50% and were identified with a minimum of 2 unique peptides

cellular growth arrest or apoptosis in EA.hy926, it can be used to study the immediate effects on the proteome.

Low-dose radiation is expected to trigger subtle proteome alterations that may be difficult to observe. To increase the sensitivity of the chosen method (2D-DIGE), the cellular proteins were fractionated and the focus was put on the cytosolic alterations. Furthermore, two overlapping pH ranges (4–7 and 6–10) in a large-size gel format were studied. In addition, two different dose rates (20 and 190 mGy/min) and two time points (4 and 24 h) were investigated. The number of protein spots detected with both pH ranges averaged 1,300. Fifteen significantly differentially expressed proteins were found, of which ten were up-regulated and five down-regulated with more than ± 1.5-fold difference compared to unexposed cells. The largest number of deregulated proteins was seen with the higher dose rate 4 h after the exposure but the number of overlapping proteins between the different conditions was 33% (5/15). Pathways influenced by the low-dose exposures included the Ran and RhoA pathways, fatty acid metabolism, and stress response. A common feature of these pathways is their regulation by the cellular redox potential.

EA.hy926 was used as a model for radiation-induced immediate vascular damage also by Sriharshan et al. [11]. Radiation-induced alterations were investigated 4 and 24 h after exposure to an acute clinically relevant dose (2.5 Gy Cs-137 gamma). This dose led to a significant growth inhibition, probably due to apoptosis. Two complementary proteomic approaches, SILAC and 2D-DIGE, were used. The whole cell proteomes of the endothelial cells were analysed 4 and 24 h after irradiation. Differentially expressed proteins were identified and quantified by MALDI-TOF/TOF and LTQ Orbitrap tandem mass spectrometry.

Using the SILAC method, a total of 3,076 unique proteins were identified at both time points. Of these, 2,572 and 2,391 were quantified at 4 and 24 h, respectively. 2,274 proteins were common to both time points. A "Christmas tree" model [12] representing normalised protein ratios of all identified proteins by SILAC plotted against summed peptide intensities is shown in Fig. 2.1. The deregulated proteins are distinguished as data points lying further away from the y-axis.

The 2D-DIGE analysis was performed using four replicates (three biological replicates and one technical replicate) at both time points. The selection criteria for deregulation were the same for both methods: p-value < 0.01 (Student's t-test), variability < 50% between biological replicates, identification based on at least two unique peptides and fold difference greater than 1.30 or less than −1.30.

Four hours after irradiation, a total of 58 proteins were differentially regulated (SILAC: 31; 2D-DIGE: 27); 12 of these were up-regulated and 47 down-regulated. The fold changes ranged

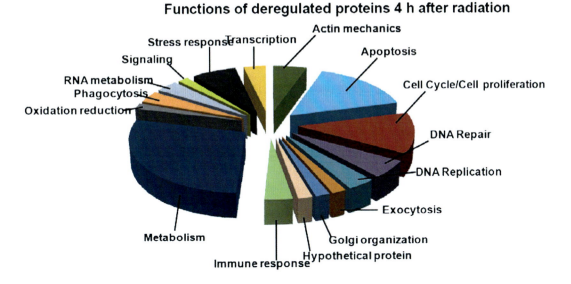

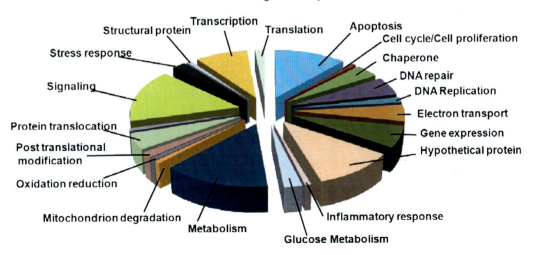

Fig. 2.2 Associated functions of proteins found to be deregulated at 4 h and 24 h after irradiation. Differentially regulated proteins were analysed for "functional categories" using the UniProt knowledge database and the PANTHER classification system

from 1.82 to −3.24. The stress response and DNA repair proteins, such as HSP90-beta, HSP75, Ku70 and Ku80, showed an immediate and transient but temporary down-regulation. Similarly, all proteins involved in metabolic processes showed immediate down-regulation.

At 24 h, a total of 136 proteins were found to be significantly deregulated (SILAC: 122; 2D-DIGE: 18). 53 were up-regulated and 83 down-regulated: several of these proteins showed more than twofold deregulation.

The functional correlation analysis of the differentially regulated proteins was done by database search using UniProt, Swiss-Prot and PANTHER. The analysis revealed several radiation-induced biological processes (Fig. 2.2). A majority (63%) of the biological processes were affected at both time points. However, DNA repair and replication, cell cycle and proliferation, stress response, apoptosis and general metabolic activity were more pronounced at 4 h. Rho-dependent rearrangement of actin cytoskeleton, immune response and alteration of the Golgi proteome were functions altered only at 4 h. In contrast, alterations in cellular signaling and transcriptional activity were more pronounced at 24 h. Glucose metabolism, inflammatory response, mitochondrial degradation and electron transport, protein translocation, post-translational modification and translational activation were processes found altered only at 24 h.

Taken together, the deregulated proteins were categorised in four key pathways: (1) glycolysis/gluconeogenesis and synthesis/degradation of ketone bodies, (2) oxidative phosphorylation, (3) Rho-mediated cell motility and (4) non-homologous end joining. Most importantly, a pronounced shift in the cellular energy metabolism including activation of the glycolytic pathway and alteration of the mitochondrial OXPHOS balance were observed.

Although there were numerous overlapping deregulated proteins found in this study, it shows the complementarity of the two proteomic methods, SILAC and 2D-DIGE, the former being more sensitive but the latter enabling the detection of protein isoforms, fragments and modified proteins.

2.2 Radiation Effects in Human Peripheral Lymphocytes

Turtoi et al. exposed human peripheral lymphocytes ex vivo by irradiating whole blood with gamma doses of 1, 2 and 4 Gy [13]. Proteomics analysis was performed using samples from one individual by 2D-PAGE 2 h after the exposure. The gels were stained with either Sypro Ruby or Pro-Q Diamond to estimate the total number of protein alterations or the number of altered phosphorylated proteins, respectively. The results were validated using immunoblotting or RT-qPCR in two additional individuals.

This study was able to find only around 20 altered proteins of which 11 could be identified. Out of five proteins showing significant down-regulation four were structural proteins (beta-actin, talin-1, talin-2, zyxin-2). Also peroxin-1 involved in protein transport and degradation was down-regulated. Four of the up-regulated proteins were associated with cell cycle control and immune system. Six proteins showed an altered phosphorylation status (major histocompatibility complex binding protein-2, phosphoglycerate kinase-1, annexin-A6, zyxin-2, interleukin-17E, beta-actin). The immunoblotting revealed large interindividual variation in the radiation responsiveness. Taken together, the study is limited by the chosen proteomics approach (2D-PAGE) that is reflected by the few alterations observed. Therefore, no extensive conclusions can be drawn from this study.

2.3 Radiation Response in the Nucleolar Proteome of Human Skin Fibroblastoid Cell Line

Moore et al. studied the effects of ionizing radiation (10 Gy of Cs-137 gamma) on the nucleoli of human WS1 skin fibroblast cell line [14]. The nucleolus is a nuclear organelle that co-ordinates rRNA transcription and ribosome subunit biogenesis but is also involved in cell cycle control, DNA damage repair and DNA processing. The nucleoli of SILAC-labelled cells were isolated 1, 3, 6, or 16 h after irradiation and compared to those from sham-irradiated control cells. Around 100 nucleolar proteins were found to be deregulated at each time point. However, the ionizing radiation (IR) treatment provoked only minor changes in the steady state content of the nucleolar proteome in comparison with UVC radiation. IR increased the nucleolar accumulation of the paraspeckle proteins at early time points after the irradiation. Non-homologous end-joining (NHEJ) proteins Ku70 and Ku80 as well as the catalytic subunit of DNA-PK (PRKDC) were detectable throughout the time course studied. Ku70/80 accumulated selectively 3 h after the exposure, whereas ILF2 and ILF3 which are dsRNA binding proteins and interact with Ku70/80 and PRKDC, decreased during the first 3 h. The level of PRKDC was fluctuating during this period. Taken together, IR did not substantially alter the nucleolar morphology and provoked highly selective, rapid proteomic responses that were fluctuating over time.

2.4 Proteome Alterations in the Irradiated Heart

2.4.1 Heart Tissue Alterations

Adverse effects of ionizing radiation on the cardiovascular system have the potential for a large impact on public health. High doses of radiation applied to the heart during radiotherapy used in breast cancer [15–18], Hodgkin's disease [19] or childhood cancers [20] increase cardiovascular incidence and mortality. Epidemiological studies indicate that much lower irradiation doses (≤ 1 Gy) typical of occupational [21–26], medical [20, 27] or environmental exposures [28, 29] also increase the risk of cardiovascular disease (CVD) several decades after the exposure. However, this remains controversial as some studies find no association between low-dose ionizing radiation and an increased risk for CVD [30–36].

Azimzadeh et al. studied the initial cardiac injury, measured as immediate changes after 4 and 24 h in the whole heart proteome of a C57BL/6 mouse exposed to total body irradiation, using a dose relevant to both accidental and intentional exposure (3 Gy gamma-ray) [37]. By using two complementary quantitative proteomic approaches, isotope coded protein labelling (ICPL) and 2D-DIGE, combined with bioinformatics analysis, several biological pathways were reported to be involved in the initial radiation-induced cardiac damage.

Pathway analyses indicated that total body irradiation immediately induced biological responses such as inflammation, antioxidative defense and reorganization of structural proteins (Fig. 2.3).

Altered levels of several members of the acute phase protein family (peroxiredoxin, hemopexin, haptoglobin, ferritin, transferrin, murinoglobulin1 and others) were found with both techniques. An immediate acute-phase response may act against the oxidative tissue damage induced by ionizing radiation [38]. In the heart tissue, active inflammation and alteration of acute phase response have been associated with the development of cardiac disease [39]. Haptoglobin and hemopexin are acute-phase proteins induced mainly by several cytokines following inflammatory processes [40]. The regulation of ferritin and transferrin, being major iron binding proteins, has been reported to be influenced by various inflammatory

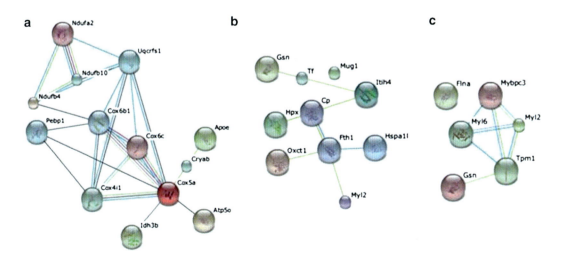

Fig. 2.3 Graphical representation of the interaction networks of differentially regulated proteins at 5 h and 24 h after total body irradiation. Association networks of the differentially expressed proteins were analyzed by the STRING 8.2 software. The main networks consisted of proteins involved in the mitochondrial respiratory activity at 24 h (**a**), acute phase response at 24 h (**b**) and cellular structural organization at 5 h (**c**)

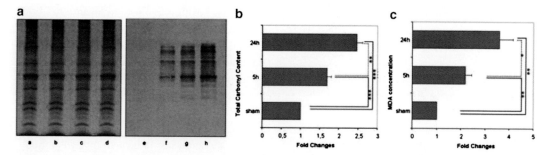

Fig. 2.4 Measurement of total protein carbonyl content and malondialdehyde (MDA) content of mouse heart lysate at 5 h/24 h after total body irradiation (TBI). The extracted proteins from sham-irradiated and irradiated hearts at 5 h and 24 h after radiation exposure were treated with 2,4-dinitrophenylhydrazine to derivatize the oxidized amino acids. SDS-PAGE- separated proteins were probed with anti-dinitrpphenol polyclonal antibody (**a**, *lanes e–h*). 10 μg of each sample was loaded per lane. Coomassie stained gel (**a**, *lanes a–d*) was used as a loading control. Higher levels of carbonylation were observed in irradiated samples compared to sham-irradiated samples (**b**). MDA concentration was measured in irradiated and sham-irradiated heart lysates using MDA assay. A significant increase of lipid peroxidation was observed after TBI (**c**) (*t* test; $^*p \leq 0.05$, $^{**}p \leq 0.005$) Three biological replicates were used in all experiments

cytokines [41]. Presumably, these proteins act in a protective manner by binding iron and thus reducing the toxicity of this metal that is a key player in the generation of reactive oxygen species (ROS) [42]. Several proteins important in the protection against oxidative damage such as glutathione transferase, complement C3, apolipoprotein E and alpha-crystallin B were also found to be up-regulated in this study.

ROS is known to cause a number of post-translational modifications of proteins, including carbonylation. Carbonylation may occur either directly through the oxidation of amino acid side chains or indirectly by addition of lipid oxidation products such as malondialdehyde to proteins [43]. The mitochondrial and structural proteins showing most alterations in the study by Azimzadeh et al. have been previously shown to be the main targets of oxygen-induced protein carbonylation [44]. The immunoblotting analysis showed that protein carbonyl content was increased in irradiated mice hearts compared to sham-irradiated ones (Fig. 2.4).

2.4.2 Alterations in the Cardiac Mitochondria

Mitochondria play a central role in oxidative metabolism, where the final products of glycolysis and fatty acid metabolism, pyruvate and acetyl CoA, are used in the Krebs cycle and by oxidative phosphorylation to produce energy. As heart tissue has a high energy demand, it is not surprising that mitochondria contribute about 40% of the total cellular volume of cardiomyocytes [45]. Approximately 90% of energy is supplied by these organelles [46]. In numerous biochemical and functional studies of cardiomyocytes, impairment of oxidative metabolism has been directly linked to the development of cardiovascular disease [47–50]. Loss of control over the reduction and oxidation processes within the mitochondria may lead to disruption of metabolic homeostasis and an increased production of ROS such as peroxide, superoxide and hydroxyl radicals. Such an excess of ROS is capable of causing damage to many cellular components including lipids, proteins, and DNA [51, 52]. Oxidative stress is also known to contribute to vascular disease and endothelial cell dysfunction potentially leading to further cardiovascular damage [53]. Conversely, lower ROS concentrations stimulate cellular signaling and gene expression modulating vascular function [54] and playing an important role in cardioprotection [55, 56].

Exposure of eukaryotic cells to radiation leads to the production of ROS within minutes. Leach et al. showed that, in the dose range between

1 and 10 Gy, the amount of ROS produced per cell was constant whereas the percentage of ROS producing cells increased with the dose [57]. This induced increase in ROS production was dependent on dysfunctional mitochondrial electron transport and was observed in several cell types.

Barjaktarovic et al. investigated whether ionizing radiation causes non-transient impairment of cardiac mitochondria that could finally lead to cardiovascular disease [58]. For this purpose C57BL/6N mice were locally irradiated to the heart using an acute dose of 0.2 Gy or 2 Gy X-ray; the control mice were sham-irradiated. The cardiac mitochondria were isolated 4 weeks after the exposure and proteomic and functional analysis was performed. Again, two complementary quantitative proteomic approaches, ICPL and 2D-DIGE, were used to investigate the radiation-induced dysregulation of protein expression.

The high dose (2 Gy) induced changes in 15 and 9 proteins, identified by ICPL or DIGE, respectively. The low dose (0.2 Gy) caused fewer alterations, the corresponding numbers of proteins showing altered expression being 5 and 1. As some of the deregulated proteins were overlapping between the methods and doses, a total of 25 mitochondrial proteins were found to be radiation-responsive. Three main biological categories were affected: the oxidative phophorylation, the pyruvate metabolism, and the cytoskeletal structure. The changes in the pyruvate metabolism and structural proteins were seen with both low and high radiation doses.

Especially affected by the radiation were the respiratory complexes I and III; the level of several subunits and the activity of these complexes were significantly downregulated after the high dose exposure. As these complexes are the main producers of mitochondrial ROS, the general level of ROS and the oxidized proteins was measured. ROS production was significantly increased leading to an enhanced oxidation of mitochondrial proteins.

This study suggests that an altered redox status will result in impaired heart function and/or increased vulnerability towards additional stress conditions in the long term. As mitochondria are closely associated with myofibrils it is possible that increased mitochondrial ROS production may lead to impaired contractility through disruption of actin-myosin interactions in the heart.

2.5 Usage of Formalin-Fixed Paraffin-Embedded Material for Proteomics

Formalin-fixed paraffin-embedded (FFPE) tissue has recently gained interest as an alternative to fresh/frozen tissue for retrospective protein biomarker discovery. In the field of radiation biology, tissue banks may provide retrospective information not only about the nature and frequency of radiation-induced disease, but also about the molecular pathways and cellular processes leading to such end-points [59]. However, during the fixation process proteins undergo degradation and cross-linking, making conventional protein analysis technologies problematic.

Azimzadeh et al. compared several extraction and separation methods for the analysis of proteins in FFPE tissues [60]. Incubation of tissue sections at high temperature with a novel extraction buffer resulted in improved protein recovery. Protein separation by 1-DE followed by LC-ESI MS/MS analysis was the most effective approach to identify proteins, based on the number of peptides reliably identified. Interestingly, a number of peptides were identified in regions of the 1DE not corresponding to their native molecular weights. This was an indication of the formation of protein-protein complexes by cross-linking, and to protein fragmentation due to prolonged sample storage.

Quantitative proteome analysis of FFPE tissue has been hampered by the lack of efficient labelling method. The usage of conventional protein labelling on FFPE has turned out to be inefficient as most labels target lysine residues that are blocked by the formalin treatment. Recently, non-labelling approaches have been suggested as an alternative for quantification of FFPE proteome profiles [61, 62].

Azimzadeh et al. applied the optimised protein extraction and separation conditions for FFPE

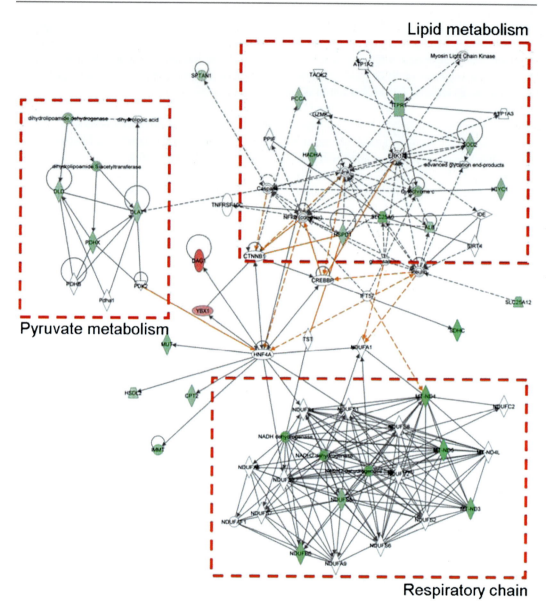

Fig. 2.5 Graphical representation of the merged most significant protein networks using Ingenuity pathway analysis. All significantly deregulated proteins were imported into the Ingenuity pathway analysis. The up-regulated proteins are marked in *red* and the down-regulated in *green*. The nodes represent proteins that are connected with one or several *arrows*; the *solid arrows* represent direct interactions and the *dotted arrows* indirect interactions; direct and predicted protein interactions were searched for (http://www.Ingenuity.com)

tissue that was previously developed for qualitative proteomics and combined it with the label-free proteomics technology [63]. As a model system they used FFPE heart tissue of control and exposed C57BL/6 mice after total body irradiation using a gamma ray dose of 3 Gy. In total, 31 significantly deregulated proteins were found in irradiated hearts 24 h after the exposure. The proteomics data were further evaluated and validated by bioinformatics and immunoblotting. The analysis indicated radiation-induced alterations in three main biological pathways: respiratory chain, lipid metabolism and pyruvate metabolism (Fig. 2.5).

The label-free approach enables the quantitative measurement of radiation-induced alterations in FFPE tissue and facilitates retrospective biomarker identification using radiobiological archives.

2.6 Protein Biomarkers of Ionizing Radiation

Assessment of health risks in humans exposed to ionizing radiation is largely based on the results from epidemiological studies. The epidemiological cohorts used until now are populations exposed to relatively high doses of radiation often given at high dose rates. Consequently, the magnitude of risk after exposure to low doses and low dose rates, as is the case in many environmental and occupational scenarios, remains controversial due to a lack of direct evidence.

It is reasonable to believe that protein expression profiling can be used to successfully find radiation-associated protein biomarkers in biological samples such as urine, blood/serum or even tissue. The primary advantage of using tissue is that protein expression is tissue-specific and some of the biomarkers will be maximally expressed in the targeted tissue. The obvious primary disadvantage is the difficulty in obtaining tissue specimens by biopsy or autopsy. In contrast, protein-rich samples of biological fluids, like serum, urine and saliva can be collected in a non-invasive or semi-invasive manner, and quantification of radiation-induced protein expression can be automated using high-throughput proteomic technologies.

However, the discovery of radiation-associated protein biomarkers remains an enormous challenge within radiobiological research because of the time- and dose-dependent variation of protein expression. Animal and cellular studies have been used as a tool to identify potential biomarkers that then may be tested in molecular epidemiological studies [7, 11, 37, 58, 64, 65]. Marchetti et al. conducted a literature review of candidate protein biomarkers for individual radiation biodosimetry of exposure to ionizing radiation [66]. The study included human, animal and cellular studies using in vivo and in vitro irradiation. All identified radiation-responsive proteins, 261 in total, including 173 human proteins, were tabulated and assigned to nine priority groups. This method resulted in a proposed panel of 20 candidate protein biomarkers for different doses and time points after exposure.

So far, only two proteins have been used as bioindicators for radiation exposure: the amylase that indicates radiation-induced damage of the parotid gland [67, 68], and Flt3-ligand, a hematopoietic cytokine indicating damage of the bone marrow [69]. Both are relatively easy to measure from serum/plasma using a clinical blood chemistry analyser or commercial sandwich enzyme immunoassay (ELISA), respectively. However, analysed separately these biomarkers are neither specific nor sensitive enough to estimate the received dose, especially if the time from the exposure exceeds 48 h.

Given the complexity of radiation response in humans, it is unrealistic to believe that a single protein marker will ever give incremental information of the exposure or etiology of the disease. A multimarker approach will be more useful as it gives information about the interplay of different possible pathways involved. The impact that good protein biomarkers will give as indicators of exposure, susceptibility or disease is more than enough to justify the use of proteomics on this area. However, whenever applicable, proteomics studies should be complemented by incorporating other technologies providing additional information about putative radiation biomarkers.

References

1. Marsden PA, Goligorsky MS, Brenner BM (1991) Endothelial cell biology in relation to current concepts of vessel wall structure and function. J Am Soc Nephrol 1(7):931–948
2. Furchgott RF, Zawadzki JV (1980) The obligatory role of endothelial cells in the relaxation of arterial smooth muscle by acetylcholine. Nature 288(5789):373–376
3. Luscher TF, Richard V, Tschudi M, Yang ZH, Boulanger C (1990) Endothelial control of vascular tone in large and small coronary arteries. J Am Coll Cardiol 15(3):519–527

4. Herrera MD, Mingorance C, Rodriguez-Rodriguez R, Sotomayor MA (2009) Endothelial dysfunction and aging: an update. Ageing Res Rev 9(2):142–152
5. Falk E, Fernandez-Ortiz A (1995) Role of thrombosis in atherosclerosis and its complications. Am J Cardiol 75(6):3B–11B
6. Ross R (1999) Atherosclerosis–an inflammatory disease. N Engl J Med 340(2):115–126
7. Pluder F, Barjaktarovic Z, Azimzadeh O, Mortl S, Kramer A, Steininger S, Sarioglu H, Leszczynski D, Nylund R, Hakanen A, Sriharshan A, Atkinson MJ, Tapio S (2011) Low-dose irradiation causes rapid alterations to the proteome of the human endothelial cell line EA.hy926. Radiat Environ Biophys 50(1):155–166. doi:10.1007/s00411-010-0342-9
8. Nylund R, Leszczynski D (2004) Proteomics analysis of human endothelial cell line EA.hy926 after exposure to GSM 900 radiation. Proteomics 4(5):1359–1365
9. Nylund R, Leszczynski D (2006) Mobile phone radiation causes changes in gene and protein expression in human endothelial cell lines and the response seems to be genome- and proteome-dependent. Proteomics 6(17):4769–4780
10. Boerma M, Burton GR, Wang J, Fink LM, McGehee RE Jr, Hauer-Jensen M (2006) Comparative expression profiling in primary and immortalized endothelial cells: changes in gene expression in response to hydroxy methylglutaryl-coenzyme a reductase inhibition. Blood Coagul Fibrinolysis 17(3):173–180
11. Sriharshan A, Boldt K, Sarioglu H, Barjaktarovic Z, Azimzadeh O, Hieber L, Zitzelsberger H, Ueffing M, Atkinson MJ, Tapio S (2012) Proteomic analysis by SILAC and 2D-DIGE reveals radiation-induced endothelial response: four key pathways. J Proteomics 75(8):2319–2330. doi:10.1016/j.jprot.2012.02.009
12. Cox J, Mann M (2008) MaxQuant enables high peptide identification rates, individualized p.p.b.-range mass accuracies and proteome-wide protein quantification. Nat Biotechnol 26(12):1367–1372. doi:10.1038/nbt.1511
13. Turtoi A, Sharan RN, Srivastava A, Schneeweiss FH (2010) Proteomic and genomic modulations induced by g-irradiation of human blood lymphocytes. Int J Radiat Biol 86(10):888–904. doi:10.3109/09553002.2010.486016
14. Moore HM, Bai B, Boisvert FM, Latonen L, Rantanen V, Simpson JC, Pepperkok R, Lamond AI, Laiho M (2011) Quantitative proteomics and dynamic imaging of the nucleolus reveal distinct responses to UV and ionizing radiation. Mol Cell Proteomics 10(10):M111 009241. doi:10.1074/mcp.M111.009241
15. Clarke M, Collins R, Darby S, Davies C, Elphinstone P, Evans E, Godwin J, Gray R, Hicks C, James S, MacKinnon E, McGale P, McHugh T, Peto R, Taylor C, Wang Y (2005) Effects of radiotherapy and of differences in the extent of surgery for early breast cancer on local recurrence and 15-year survival: an overview of the randomised trials. Lancet 366(9503):2087–2106
16. Darby S, Hill D, Auvinen A, Barros-Dios JM, Baysson H, Bochicchio F, Deo H, Falk R, Forastiere F, Hakama M, Heid I, Kreienbrock L, Kreuzer M, Lagarde F, Makelainen I, Muirhead C, Oberaigner W, Pershagen G, Ruano-Ravina A, Ruosteenoja E, Rosario AS, Tirmarche M, Tomasek L, Whitley E, Wichmann HE, Doll R (2005) Radon in homes and risk of lung cancer: collaborative analysis of individual data from 13 European case–control studies. BMJ 330(7485):223
17. Darby S, McGale P, Peto R, Granath F, Hall P, Ekbom A (2003) Mortality from cardiovascular disease more than 10 years after radiotherapy for breast cancer: nationwide cohort study of 90,000 Swedish women. BMJ 326(7383):256–257
18. Demirci S, Nam J, Hubbs JL, Nguyen T, Marks LB (2009) Radiation-induced cardiac toxicity after therapy for breast cancer: interaction between treatment era and follow-up duration. Int J Radiat Oncol Biol Phys 73(4):980–987
19. Swerdlow AJ, Higgins CD, Smith P, Cunningham D, Hancock BW, Horwich A, Hoskin PJ, Lister A, Radford JA, Rohatiner AZ, Linch DC (2007) Myocardial infarction mortality risk after treatment for Hodgkin disease: a collaborative British cohort study. J Natl Cancer Inst 99(3):206–214
20. Tukenova M, Guibout C, Oberlin O, Doyon F, Mousannif A, Haddy N, Guerin S, Pacquement H, Aouba A, Hawkins M, Winter D, Bourhis J, Lefkopoulos D, Diallo I, de Vathaire F (2010) Role of cancer treatment in long-term overall and cardiovascular mortality after childhood cancer. J Clin Oncol 28(8):1308–1315
21. Azizova TV, Muirhead CR, Druzhinina MB, Grigoryeva ES, Vlasenko EV, Sumina MV, OHJ A, Zhang W, Haylock RGE, Hunter N (2010) Cardiovascular diseases in the cohort of workers first employed at Mayak PA in 1948–1958. Radiat Res 174(2):155–168
22. Hauptmann M, Mohan AK, Doody MM, Linet MS, Mabuchi K (2003) Mortality from diseases of the circulatory system in radiologic technologists in the United States. Am J Epidemiol 157(3):239–248
23. Howe GR, Zablotska LB, Fix JJ, Egel J, Buchanan J (2004) Analysis of the mortality experience amongst U.S. nuclear power industry workers after chronic low-dose exposure to ionizing radiation. Radiat Res 162(5):517–526
24. Ivanov VK (2007) Late cancer and noncancer risks among Chernobyl emergency workers of Russia. Health Phys 93(5):470–479
25. McGeoghegan D, Binks K, Gillies M, Jones S, Whaley S (2008) The non-cancer mortality experience of male workers at British Nuclear Fuels plc, 1946–2005. Int J Epidemiol 37(3):506–518
26. Muirhead CR, O'Hagan JA, Haylock RG, Phillipson MA, Willcock T, Berridge GL, Zhang W (2009) Mortality and cancer incidence following occupational radiation exposure: third analysis of the national registry for radiation workers. Br J Cancer 100(1):206–212

27. Carr ZA, Land CE, Kleinerman RA, Weinstock RW, Stovall M, Griem ML, Mabuchi K (2005) Coronary heart disease after radiotherapy for peptic ulcer disease. Int J Radiat Oncol Biol Phys 61(3):842–850
28. Preston DL, Shimizu Y, Pierce DA, Suyama A, Mabuchi K (2003) Studies of mortality of atomic bomb survivors. Report 13: solid cancer and non-cancer disease mortality: 1950–1997. Radiat Res 160(4):381–407
29. Shimizu Y, Kodama K, Nishi N, Kasagi F, Suyama A, Soda M, Grant EJ, Sugiyama H, Sakata R, Moriwaki H, Hayashi M, Konda M, Shore RE (2010) Radiation exposure and circulatory disease risk: Hiroshima and Nagasaki atomic bomb survivor data, 1950–2003. BMJ 340:b5349
30. Darby SC, Doll R, Gill SK, Smith PG (1987) Long term mortality after a single treatment course with X-rays in patients treated for ankylosing spondylitis. Br J Cancer 55(2):179–190
31. Davis FG, Boice JD Jr, Hrubec Z, Monson RR (1989) Cancer mortality in a radiation-exposed cohort of Massachusetts tuberculosis patients. Cancer Res 49(21):6130–6136
32. Kreuzer M, Grosche B, Schnelzer M, Tschense A, Dufey F, Walsh L (2010) Radon and risk of death from cancer and cardiovascular diseases in the German uranium miners cohort study: follow-up 1946–2003. Radiat Environ Biophys 49(2):177–185
33. Kreuzer M, Kreisheimer M, Kandel M, Schnelzer M, Tschense A, Grosche B (2006) Mortality from cardiovascular diseases in the German uranium miners cohort study, 1946–1998. Radiat Environ Biophys 45(3):159–166
34. Little MP, Tawn EJ, Tzoulaki I, Wakeford R, Hildebrandt G, Paris F, Tapio S, Elliott P (2010) Review and meta-analysis of epidemiological associations between low/moderate doses of ionizing radiation and circulatory disease risks, and their possible mechanisms. Radiat Environ Biophys 49(2):139–153
35. Richardson DB, Wing S (1999) Radiation and mortality of workers at Oak Ridge National Laboratory: positive associations for doses received at older ages. Environ Health Perspect 107(8):649–656
36. Talbott EO, Youk AO, McHugh-Pemu KP, Zborowski JV (2003) Long-term follow-up of the residents of the three mile Island accident area: 1979–1998. Environ Health Perspect 111(3):341–348
37. Azimzadeh O, Scherthan H, Sarioglu H, Barjaktarovic Z, Conrad M, Vogt A, Calzada-Wack J, Neff F, Aubele M, Buske C, Atkinson MJ, Tapio S (2011) Rapid proteomic remodeling of cardiac tissue caused by total body ionizing radiation. Proteomics 11(16):3299–3311. doi:10.1002/pmic.201100178
38. Chen C, Lorimore SA, Evans CA, Whetton AD, Wright EG (2005) A proteomic analysis of murine bone marrow and its response to ionizing radiation. Proteomics 5(16):4254–4263
39. Oldgren J, Wallentin L, Grip L, Linder R, Norgaard BL, Siegbahn A (2003) Myocardial damage, inflammation and thrombin inhibition in unstable coronary artery disease. Eur Heart J 24(1):86–93
40. Bowman BH, Kurosky A (1982) Haptoglobin: the evolutionary product of duplication, unequal crossing over, and point mutation. Adv Hum Genet 12(189–261):453–184
41. Rogers JT, Bridges KR, Durmowicz GP, Glass J, Auron PE, Munro HN (1990) Translational control during the acute phase response. Ferritin synthesis in response to interleukin-1. J Biol Chem 265(24):14572–14578
42. Iwai K, Drake SK, Wehr NB, Weissman AM, LaVaute T, Minato N, Klausner RD, Levine RL, Rouault TA (1998) Iron-dependent oxidation, ubiquitination, and degradation of iron regulatory protein 2: implications for degradation of oxidized proteins. Proc Natl Acad Sci U S A 95(9):4924–4928
43. Madian AG, Regnier FE (2010) Proteomic identification of carbonylated proteins and their oxidation sites. J Proteome Res 9(8):3766–3780
44. Khaliulin I, Schneider A, Houminer E, Borman JB, Schwalb H (2004) Apomorphine prevents myocardial ischemia/reperfusion-induced oxidative stress in the rat heart. Free Radic Biol Med 37(7):969–976
45. White MY, Edwards AV, Cordwell SJ, Van Eyk JE (2008) Mitochondria: a mirror into cellular dysfunction in heart disease. Proteomics Clin Appl 2(6):845–861. doi:10.1002/prca.200780135
46. Schlattner U, Tokarska-Schlattner M, Wallimann T (2006) Mitochondrial creatine kinase in human health and disease. Biochim Biophys Acta 1762(2):164–180
47. Ballinger SW (2005) Mitochondrial dysfunction in cardiovascular disease. Free Radic Biol Med 38(10):1278–1295
48. Dhalla NS, Temsah RM, Netticadan T (2000) Role of oxidative stress in cardiovascular diseases. J Hypertens 18(6):655–673
49. Misra MK, Sarwat M, Bhakuni P, Tuteja R, Tuteja N (2009) Oxidative stress and ischemic myocardial syndromes. Med Sci Monit 15(10):RA209–RA219
50. Takano H, Zou Y, Hasegawa H, Akazawa H, Nagai T, Komuro I (2003) Oxidative stress-induced signal transduction pathways in cardiac myocytes: involvement of ROS in heart diseases. Antioxid Redox Signal 5(6):789–794
51. Allan IM, Vaughan AT, Milner AE, Lunec J, Bacon PA (1988) Structural damage to lymphocyte nuclei by H2O2 or gamma irradiation is dependent on the mechanism of OH. radical production. Br J Cancer 58(1):34–37
52. Eny KM, El-Sohemy A, Cornelis MC, Sung YK, Bae SC (2005) Catalase and PPARgamma2 genotype and risk of systemic lupus erythematosus in Koreans. Lupus 14(5):351–355

53. Coyle CH, Kader KN (2007) Mechanisms of H2O2-induced oxidative stress in endothelial cells exposed to physiologic shear stress. ASAIO J 53(1):17–22
54. Wolf G (2000) Free radical production and angiotensin. Curr Hypertens Rep 2(2):167–173
55. Garlid KD, Costa AD, Quinlan CL, Pierre SV, Dos Santos P (2009) Cardioprotective signaling to mitochondria. J Mol Cell Cardiol 46(6):858–866
56. Hausenloy DJ, Yellon DM (2008) Preconditioning and postconditioning: new strategies for cardioprotection. Diabetes Obes Metab 10(6):451–459
57. Leach JK, Van Tuyle G, Lin PS, Schmidt-Ullrich R, Mikkelsen RB (2001) Ionizing radiation-induced, mitochondria-dependent generation of reactive oxygen/nitrogen. Cancer Res 61(10):3894–3901
58. Barjaktarovic Z, Schmaltz D, Shyla A, Azimzadeh O, Schulz S, Haagen J, Dörr W, Sarioglu H, Schäfer A, Atkinson MJ, Zischka H, Tapio S (2011) Radiation–induced signaling results in mitochondrial impairment in mouse heart at 4 weeks after exposure to X-rays. PLoS One 6(12):e27811. doi:10.1371/journal.pone.0027811
59. Tapio S, Atkinson MJ (2008) Molecular information obtained from radiobiological tissue archives: achievements of the past and visions of the future. Radiat Environ Biophys 47(2):183–187
60. Azimzadeh O, Barjaktarovic Z, Aubele M, Calzada-Wack J, Sarioglu H, Atkinson MJ, Tapio S (2010) Formalin-fixed paraffin-embedded (FFPE) proteome analysis using gel-free and gel-based proteomics. J Proteome Res 9(9):4710–4720. doi:10.1021/pr1004168
61. Donadio E, Giusti L, Cetani F, Da Valle Y, Ciregia F, Giannaccini G, Pardi E, Saponaro F, Torregrossa L, Basolo F, Marcocci C, Lucacchini A (2011) Evaluation of formalin-fixed paraffin-embedded tissues in the proteomic analysis of parathyroid glands. Proteome Sci 9(1):29. doi:10.1186/1477-5956-9-29
62. Ostasiewicz P, Zielinska DF, Mann M, Wisniewski JR (2010) Proteome, phosphoproteome, and N-glycoproteome are quantitatively preserved in formalin-fixed paraffin-embedded tissue and analyzable by high-resolution mass spectrometry. J Proteome Res 9(7):3688–3700. doi:10.1021/pr100234w
63. Azimzadeh O, Scherthan H, Yentrapalli R, Barjaktarovic Z, Ueffing M, Conrad M, Neff F, Calzada-Wack J, Aubele M, Buske C, Atkinson MJ, Hauck SM, Tapio S (2012) Label-free protein profiling of formalin-fixed paraffin-embedded (FFPE) heart tissue reveals immediate mitochondrial impairment after ionising radiation. J Proteomics 75(8):2384–2395. doi:10.1016/j.jprot.2012.02.019
64. Guipaud O, Holler V, Buard V, Tarlet G, Royer N, Vinh J, Benderitter M (2007) Time-course analysis of mouse serum proteome changes following exposure of the skin to ionizing radiation. Proteomics 7(21):3992–4002
65. Tapio S, Danescu-Mayer J, Asmuss M, Posch A, Gomolka M, Hornhardt S (2005) Combined effects of gamma radiation and arsenite on the proteome of human TK6 lymphoblastoid cells. Mutat Res 581(1–2):141–152
66. Marchetti F, Coleman MA, Jones IM, Wyrobek AJ (2006) Candidate protein biodosimeters of human exposure to ionizing radiation. Int J Radiat Biol 82(9):605–639. doi:10.1080/09553000600930103
67. Balzi M, Cremonini D, Tomassi I, Becciolini A, Giannardi G, Pelu G (1979) Radiation effects on the parotid gland of mammals. Part 2: modifications of plasma and parotid amylase activity. Strahlentherapie 155(8):566–569
68. Barrett A, Jacobs A, Kohn J, Raymond J, Powles RL (1982) Changes in serum amylase and its isoenzymes after whole body irradiation. Br Med J (Clin Res Ed) 285(6336):170–171
69. Bertho JM, Demarquay C, Frick J, Joubert C, Arenales S, Jacquet N, Sorokine-Durm I, Chau Q, Lopez M, Aigueperse J, Gorin NC, Gourmelon P (2001) Level of Flt3-ligand in plasma: a possible new bio-indicator for radiation-induced aplasia. Int J Radiat Biol 77(6):703–712. doi:10.1080/09553000110043711

Radiation Treatment Effects on the Proteome of the Tumour Microenvironment

Michael J. Atkinson

Abstract

Exposure of tumourous tissue to ionizing radiation initiates a wound-healing response involving remodelling of the extracellular microenvironment. The initial reaction involves direct damage to the matrix proteins and the secretion and activation of proteolytic enzymes that lead to local destruction of the extracellular matrix. Subsequently the wounded area may undergo complete repair, may enter a prolonged period of heightened proteolysis, or may overproduce matrix proteins leading to fibrosis. The source of matrix degrading enzymatic activity may be the tumour cells and the tumour stroma. Additional complexity is provided by proteolytic activity released from tissue macrophages, mast cells and by invading inflammatory cells. The local production of growth factors, including VEGF and TGF-β play a key role in coordinating the response. It is anticipated that the application of modern proteomic technologies will reveal hitherto unrecognised levels of complexity in these processes. Hopefully this will lead to the development of new therapeutic strategies to prevent long-term health implications of radiation exposure.

Keywords

Extracellular matrix • ECM • Tumour microenvironment • Ionizing radiation • Radiation injury • Inflammation • Hypoxia • Metastasis • Bystander effect • Cytokine • Lysis • Proteolysis • Stroma • Signaling pathways

3.1 The Tumour Microenvironment

The extracellular milieu of a tumour is established by the juxtaposition of tumour, non-tumour cells within the tumour (tumour stroma), the surrounding normal tissues, and the extracellular

M.J. Atkinson (✉)
Helmholtz Zentrum München, German Research Centre for Environmental Health, Institute of Radiation Biology, Ingolstaedter Landstrasse 1, Neuherberg 85764, Germany
e-mail: atkinson@helmholtz-muenchen.de

matrix. This unique biological microenvironment fulfils both a structural and regulatory role. The complexity of the tumour interface is continually shifting as the tumour progresses. These changes are initiated by loss of a differentiated tumour phenotype, the assumption of a more infiltrative or metastatic behaviour, and by on-going local and systemic responses to the tumour. They may also be brought about by anti-cancer interventions, including radiation therapy. In each instance the alterations are the net result of changes to all of the different components contributing to the microenvironment.

3.1.1 The Cellular Component of the Microenvironment

Tumour cells influence the tumour microenvironment through the production and secretion of matrix proteins, metabolites, and signalling molecules/growth factors. Tumour cells are also sources of extracellular proteases and protease-regulating inhibitors capable of remodelling the extracellular matrix.

A range of non-tumour cells may populate tumours. These include fibroblasts, vascular endothelial cells and components of the immune system. This "tumour stroma" arises as adjacent non tumour cells are engulfed by a rapidly growing tumour, are recruited in response to signalling from the tumour, or as immune cells enter the tumour as part of an inflammatory response. The diversity of the stromal cells types increases the complexity of the extracellular environment through their own secreted proteins and through interactions mediated by cell-cell and cell-matrix recognition molecules.

Compression of adjacent tissues, release of cell and matrix degradation products and the secretion of growth factors all contribute to the local inflammatory response, involving both innate and active immune responses [1]. In addition to activation of resident mast cells and tissue macrophages, the tumour may be infiltrated by mononuclear cells, such as lymphocytes, neutrophils and macrophages. The cytokine profile of the tumour microenvironment reflects this infiltration.

3.1.2 The Extracellular Matrix

The extracellular matrix (ECM) is a complex protein network made up of a collagen-mixture scaffold, elastin, proteoglycans including decorin, aggrecan and syndecan and glycoproteins such as tenascin, vitronectin, laminin, thrombospondin and fibronectin. The exact composition is dependent upon the tissue type and function. The bone matrix represents a special case as this includes a mineralized component, hydoxyapatite. Type I collagen represents the most abundant ECM component. A number of biologically active proteins regulating cell dynamics and interactions co-locate to the matrix, including growth factors and cytokines. Local matrix destruction via tumour-mediated proteolysis may be an essential component of neoangiogenesis, but loss of contact is also a key step in the attainment of a metastatic or infiltrative growth phenotype by the tumour.

3.1.3 Adhesion of Tumour Cells to the ECM

Cell surface adhesion molecules of the integrin family of integral transmembrane proteins serve to anchor tumour and stromal cells to ECM components. The extracellular ligand-binding domain engages specific motives in the ECM whilst the intracellular domain affixes the integrin to the cytoskeleton and signalling complexes. The integrins are heterodimeric proteins made up of one alpha and one beta subunit. A set of 24 possible permutations between the different alpha and beta subunits creates a range of possible integrin-matrix protein interactions.

In addition to maintaining structural integrity the integrins are involved in intracellular signalling, presumably to signal the cell-matrix interactions. Signalling is achieved through lateral mobility of ligand-bound integrins within the plasma membrane, creating islands (focal adhesions). These islands are docking sites for non-receptor tyrosine kinases (FAK, SRC) and accessory molecules capable of triggering intracellular pathways such as the GTPase/Rho, PI3/AKT and ERK pathways. The adhesion of cells within a tissue is mediated by the cadherin family of adhesion molecules responsible for homotypic cell-cell contacts.

A loss of matrix-integrin connectivity results in the rapid death of normally adherent non-transformed cells via an apoptosis-like cell death (Anoikis). Malignant cells are insensitive, having acquired the ability to survive detachment from both matrix and from adjacent cells. Additional changes in the expression of cell-cell adhesion molecules, acquisition of enhanced motility and the ability to lyse components of the extracellular matrix may all promote the detachment of single cells out of a tissue.

3.1.4 Remodelling of the Extracellular Matrix

The ECM is a dynamic structure, undergoing remodelling through degradation by proteolytic enzymes released as latent enzymes by both tumour and stromal elements. Matrix degradation is strictly controlled by the activation of these proteases and removal of endogenous protease inhibitors (see review by Queseda et al. [2]). The range of substrate specificities and modes of activation involved in ECM proteolysis are subject of a recent review by Mason [3]. Of these, the matrix metalloproteinase (MMPs) and the plasminogen activators (PAs) are those most closely involved in radiation-induced changes to the ECM.

The 27 recognised members of the matrix metalloproteinase family of zinc-dependent enzymes are responsible for removal of all of the structurally relevant proteins. Secreted MMPs are not constitutively active. They require both enzymatic cleavage and removal of non-covalently bound inhibitory proteins (tissue-inhibitors of MMPs—TIMPs). Local regulation of the balance between MMPs and TIMPs is determined by control of their synthesis and secretion, most prominently by cytokines and inflammatory mediators.

The urokinase-like plasminogen activator (uPA) is a protease that, upon capture at cell surfaces by a specific uPA receptor, is able to locally convert plasminogen into the active protease plasmin. In turn plasmin is able to initiate further proteolytic cascades by cleaving latent proteases such as the Matrix metalloproteinase (MMPS). uPA activity is regulated by plasminogen activator inhibitors (PA-Is), and it is the balance between receptor, uPA, and inhibitors that determine the net proteolytic activity [4].

3.1.5 Vascularization and Remodelling of the ECM Under Hypoxic Conditions

Rapid tumour growth can outpace the capacity of local blood vessels to maintain adequate oxygenation of the tumour. The resulting regions of local hypoxia have considerable consequences for tumour therapy. Not only are chemotherapeutic agents denied entry into avascular areas, but hypoxic cells are inherently radioresistant. More recently it has been suggested that hypoxia promotes dedifferentiation of surrounding stromal cells, creating a niche capable of supporting tumour repopulating stem cells [5].

There is a qualitative difference in the structure of Type 1 Collagen fibres in the extracellular matrix between normoxic and hypoxic regions of experimental tumours. In hypoxic areas the collagen fibres are less structured, presumably due to stimulation of local proteolytic activity [6] or reduced synthesis of collagen cross-links [7]. These change could facilitate entry of new blood vessels through the more open ECM architecture or even promote the transmission of angiogenic factors to neighbouring vessels [8]. Hypoxia is able to modulate the integrin-matrix interaction in an osteosarcoma cell line, promoting adhesion of cells to matrix through enhanced integrin-fibronectin binding [9].

In epithelial tumours hypoxic conditions may promote a unique pathophysiological process, the epithelial-mesenchymal transition (EMT) [10]. The EMT reiterates changes seen during the early histiogenesis of the embryo, and confers on carcinoma cells a shift in their adhesion and matrix synthesis. This results in acquisition of the capacity to conduct extensive lysis of both basement membranes and the proteins of the ECM and is a key step in epithelial tumour progression (see review by Thiery [11]).

Angiogenesis, the formation of new blood vessels from outgrowths of existing vessels, represents a critical stage in tumour development. Growth limitations due to local tissue hypoxia are overcome by the formation of blood vessels in the tumour tissue. In the absence of oxygen the tumour cells and surrounding stroma are induced to secrete pro-angiogenic factors, including vascular endothelial growth factor (VEGF) into the ECM. The latter serves to stimulate endothelial cell growth along the VEGF gradient from nearby vessels into the hypoxic areas.

Rapid tumour growth may quickly exceed the capacity of the ECM to stretch to accommodate the expansion of the tumour volume, increasing local mechanical pressure. At the same time the incomplete vascularization of the tumour may cause a local increase in hydrostatic pressure in those blood vessels close to the tumour. A direct consequences of such local physical stresses is the up-regulation of matrix production (see review by Jean et al. [12]).

3.2 Radiation Effects on the Microenvironment

The formation of ionisation-induced reactive radicals along the radiation track is considered to be the primary cause of the damage created within irradiated cells. Through radicle damage to macromolecules, in particular DNA, the effects of radiation exposure may overwhelm cellular repair capacity, leading to cell killing.

As a consequence of the direct cell killing there is a risk of inducing long-term damage in non-target tissues. Indeed, the radiation damage to adjacent normal tissue, in particular the highly radiosensitive endothelial cells of the vasculature, is a major limiting factor for the doses that can be applied during tumour radiotherapy. If the field of damage is extensive enough there may be macroscopically visible disturbances in the healing processes.

A more insidious form of damage may develop in tissues receiving lower, non-lethal, doses. Although no acute damage is apparent long-term health effects may develop, in which chronic tissue injury develops. Typical of these are the cognitive impairments, cardiovascular diseases, cataracts and diabetes, which are recorded in some survivors of childhood cancer who received radiotherapy and in survivors of the atomic bombings.

3.2.1 Alteration in ECM During Acute Phase of Damage

In cancer patients treated by surgery and radiation the wound healing response of surgical incisions is delayed. This is primarily due to a suppression of fibrogenesis at the wound site (see review by [13]). Thus, grafting of non-irradiated tissue onto irradiated sites is impaired by pre-irradiation of the graft bed [14]. Replenishment of non-irradiated fibroblasts to the site can overcome the healing deficiency [15]. In vivo studies using irradiated mice reveal a rapid decline in the level of proteoglycans within the basal lamina [16]. A reduction in the production of fibronectin is also seen immediately in irradiated fibroblasts, suggested loss of ECM production or matrix degradation contribute to the reduced healing capacity [17]. Indeed, increased expression of matrix degradation products can be correlated with production of local mediators such as TGF-β and cytokines at the wound sites [14, 18, 19].

In head and neck cancer patients with chronic open wounds post radiotherapy the tissue expression of MMP2 and MMP9 were elevated, compared to normal skin samples [20]. It is suggested that this sustained proteolytic activity, rather than hypoxia or inflammation is responsible for the long-term failure of radiation wounds to heal.

The shift between the acute phase or matrix suppression and late phases of hyperproduction may repeat itself in waves or cycles of damage repair frequently seen in severe radiation injuries. A role for the persistent inflammation in driving the cycling has been suggested [21].

3.2.2 Direct Damage to the ECM by Radiation

It has been suggested that a direct destruction of bone and skin matrix collagen can be produced by tissue irradiation, with the dose dependent release of cross-linked collagen degradation products occurring in porcine cadaver tissue maintained at 4C [22] and in dental pulp of extracted human molars [23]. A study of bone matrix using in situ Fourier transform infrared microscopy did not detect collagen degradation per se, but did detect decarboxylation of collagen protein side chains, which may be a mechanism for the radiation-induced changes in the inorganic matrix in bone [24]. Whatever the biochemical events are, irradiated extracellular matrix is sufficiently changed to activate remodeling activity. Repopulation of previously irradiated artificial tissue matrix (AlloDerm) was accompanied by an increased matrix proteolysis and activation of new matrix synthesis [25]. Thus, the direct damage of matrix macromolecules must be considered as a potential trigger for the hyperproduction of matrix following irradiation.

3.2.3 Long-Term Changes to the ECM Following Irradiation

Almost paradoxically, the later stages of the radiation injury repair, which may take days-months to appear, involves a "reactive phase" with overproduction of collagen by activated fibroblasts at the site of damage. This fibrotic reaction appears to be a rebound effect with the overproduction of matrix continuing long after the initial damage response. Irradiated mouse mammary gland revealed increased Collagen III, laminin and tenascin levels within 1 day of exposure [26]. In mice the excessive production of type I, III and IV collagens in irradiated lung was sustained over a 20–30 week period post irradiation [27]. The overproduction of fibronectin and laminin was sustained for at last 60 weeks [28] In both studies, the excessive matrix overproduction slowly declined to near-normal levels. The shift between acute damage and regeneration phases is seen in the chick chorioallantoic membrane following irradiation, with down-regulation of both protein and mRNA for fibronectin, laminin, collagen type I, integrin alpha (v)- beta3 and MMP-2 occurring after 6 h, only to increase back above initial levels by 24 h [29].

A study of transcribed genes in intestinal tissue from patients with radiation enteritis revealed excess activity of matrix protein genes, matrix remodelling genes and inflammatory response mediators [30]. This transcriptional activation results in increased synthesis of collagen matrix by the tissues [31]. Blockade of cytoskeletal interaction with the Rho GTPases [31, 32], or an anti-inflammatory intervention [33] limit excessive production of matrix in animal models of gut and lung radiation-induced fibrosis respectively.

Potential limitation of remodelling by suppressing the uPA-mediated activation of proteases is seen in biopsies of rectal tissue following radiation therapy for carcinoma, where PAI1 is up-regulated at the same time as MMP2 [34]. Indeed, a similar radiation-induced up-regulation of PAI-1 is seen in irradiated kidney epithelial cells [35] where an up-regulation of type I collagen production also occurs [36].

3.2.4 Radiation Induced Changes to the Bone Microenvironment

The extracellular matrix of the skeleton represents a special case, due to the regulated remodelling and high complexity of the non-mineralized and mineralized matrix. Radiation induced changes are shaped by the extent of cell killing, leading to replacement of original bone matrix by a more fracture-prone bone or even to radiation-induced necrosis in extreme situations [37]. In situations where the skeletal integrity has been disrupted prior to irradiation (e.g. lytic tumour foci or fractures) the rate of healing is impaired [38]. Loss of structural elements in matrix may result from direct degradation of macromolecules in the matrix [22]. In vitro studies show that the response of osteoblastic cells to radiation is similar to fibroblasts. After an initial down-regulation of the differentiated phenotype, the reacquisition of matrix synthesis results in increased secretion of matrix and eventually excessive mineralization [39], reminiscent of the fibrosis response in soft tissues. Indeed, production of TGF-β, which is postulated as the coordinator of tissue fibrosis after radiation, is up-regulated in irradiated osteoblastic cells in vitro [40].

3.2.5 Inflammation and Cellular Infiltration

The response to irradiation includes infiltration of the site by mononuclear leukocytes [41], resulting in changes to the ECM due to local proteolysis and changes in signalling due to cytokine release [42]. Thus, in vitro irradiation of a macrophage cell line induced increased MMP9 expression, without concomitant up-regulation of the inhibitory TIMP1 protein [43]. The ubiquitous tissue mast cells are also stimulated by radiation to release biologically active proteins in vitro (TNF-alpha, tryptase) [44].

3.2.6 Hypoxia and Neovascularization

Radiation may directly influence VEFG production, as increased release is reported from both cultured glioblastoma cells and chicken embryo chorioallantoic membrane exposed to X-irradiation [45, 46]. Local irradiation of the vascularized chick chorioallantoic membrane increased the level of degradation products from type IV collagen, presumably as a result of local remodelling activation [47]. As collagen fragments are one of a number of matrix degradation products with ascribed angiogenic-stimulating potential it is probably that local matrix damage by radiation may also trigger angiogenesis [46]. Indeed, addition of tumour cells to the irradiated chick membranes resulted in enhanced vascularization [46], suggesting both matrix breakdown products and VEGF secretion may contribute to local angiogenesis in irradiated tissues.

During angiogenesis the developing blood vessel must move through the extracellular matrix. In part local proteolysis is responsible for the penetration, with MMPs, plasmin and plasminogen activators all being present at the endothelial cell—matrix interface [48]. Irradiation of developing vessels in the chick chorioallantoic membrane model cause transient loss of both matrix and protease production, but this is followed within 48 h by an overexpression of fibronectin, laminin, type I collagen, MMP2 and MMP6 [29]. In the human rectum a similar phenomenon takes place, with increased vascularization accompanying fibrosis, but here the individual components differ. Thus, unlike the chick model, MMP8 and uPA appear to be the primary matrix degrading activities, whilst angiogenin and not VEGF activate vessel cell proliferation [49]. A major difference between the two models is the contribution of inflammatory cells. In the rectum the expression of both uPA and MMP8 comes from infiltrating leukocytes and not from endogenous tissue components [49].

3.2.7 Lysis of Matrix Components Following Irradiation

Studies of post-irradiated cells and tissues consistently show increased expression of proteolytic enzymes directed against ECM components. Thus, in irradiated mouse lung tissue a shift in the balance of MMP/TIMP occurs several weeks after irradiation, with increased levels of MMP2 and MMP9 that target Collagen type IV and elastin. Although apparently paradoxical, the same study also detected inhibition of MMP9, mediated through an up-regulation of TIMP expression [50]. The author's plausible explanation is that the initial activation of matrix lysis during the acute damage response gives way to a later phase of reduced matrix lysis, leading to accumulation of matrix components and the development of fibrosis.

Comparable changes in matrix remodelling were reported in cultured rat renal tubular epithelial cells, with PAI-I and MMP2 production increased [36]. Similar radiation-induced increases in MMP2 activity are also reported in rodent brain, where an increased degradation of type IV collagen is suggested to be the final consequence [51]. In in vitro irradiated human lens epithelial cells the same transient up-regulation of MMP2 and MMP9 transcription and protein levels were detected within hours of irradiation [52]. The unique biological activities of lens epithelial cells and the lens matrix do not allow a generalization of this observation, but it is possible that MMP-mediated remodelling of the lens tissue matrix may influence cataract formation after exposure to ionizing radiation.

The genetic absence of the plasminogen activator inhibitor PAI-1, or the pharmacological inhibition of PAI-I action, both ameliorated the acute radiation injury to mouse intestine [53]. This indicates that plasminogen activation of downstream proteases such as plasmin, cathepsins and MMPs is an essential regulatory component limiting the radiation matrix lysis.

3.2.8 Promotion of Metastatic Potential by Radiation-Induced Modification to the ECM

The radiation-induced increases in matrix lysis are suspected of increasing the metastatic capacity of tumour cells surviving a radiotherapy regimen. Thus, in a glioblastoma cell line irradiation promoted metastatic behaviour that could be reversed by knock down of MMP9 gene expression [54]. The ability of irradiated hepatocellular carcinoma cells to invade in vitro was enhanced following irradiation. This was accompanied by an up-regulation of MMP9 [55]. Inhibition of MMP9, either pharmacologically or by gene silencing, restored a non-invasive phenotype [55]. Interestingly, the activation of MMP9 expression by radiation in hepatocellular carcinoma cells (Cheng JC 2006) and of MMP2 in glioblastoma cells [56] was mediated by activation of the PI3kinase/AKT pathway. This may be a triggered by hepatocyte growth factor (HGF) and its receptor cMET, both being up-regulated by irradiation of neuroblastoma cells, and this correlate with increased matrix degradation and metastatic potential [57]. Interestingly, given the ability of cells to enrich surface urokinase (uPA) activity through capture by expressed cell surface uPA receptor, a study of rectal tissue following radiotherapy revealed an increase in uPA on adjacent stromal tissue, with the probable source being the irradiated tumour [34]. Such changes would promote infiltrative growth of an expanding tumour into the surrounding stroma [4].

3.2.9 Radiation-Exposure Promoting Microenvironment Changes Though Epithelial-Mesenchymal Transition

The epithelial-mesenchymal transition EMT is accompanied by acquisition of a more metastatic phenotype due to changes in adhesion potential and expression of proteolytic activities. An early observation of possible EMT induction after irradiation was provided by the work of Mary Helen Barcellos-Hoff, who observed loss of epithelial markers, adhesion and matrix interactions in irradiated mammary cells in vitro [58]. The changes recorded are similar to those that are now considered to typify the EMT and are suggested to represent irradiated tissue responses to wounding [59]. The changes of the EMT following radiation exposure are dependent upon increased TGF-β synthesis [60].

3.2.10 Changes in the Tumour Cell-Matrix Interactions Following Irradiation

The local contacts between tumour cells and their extracellular matrix is influenced by the cellular responses to the radiation exposure, and in turn can influence cell responses to the radiation [61]. For example, bovine endothelial cells grown in an artificial extracellular matrix show better survival following irradiation than cells grown in a matrix poor environment [62]. A key change following irradiation is a reduction in integrin-mediated cell-matrix binding, along with the resultant changes in intracellular signalling pathways. Thus radiation down-regulates cell surface expression of the integrin adhesion complex in prostate carcinoma cells [63].

Blockade of the integrin β1 chain by specific antibody prevented cell adhesion to a laminin-enriched synthetic matrix and sensitized cells to radiation-induced apoptosis [64]. Similarly, the siRNA-mediated silencing of either αv-β3 or αv-β5 integrins reduced glioblastoma survival after irradiation [65], as did interference with integrin-fibronectin association in mammary cultures [66]. The interaction between integrins and matrix components may be sufficient to influence survival of cancer patients receiving radiotherapy, as high integrin levels correlated with reduced patient survival [66].

3.2.11 Radiation-Induced Changes in Cell-Cell Signalling Within the Tumour Microenvironment

Transforming growth factor-β is a key player in coordinating changes in the proteolytic and adhesive activities of tumour cells and matrix. The secretion of TGF-β into the microenvironment is up-regulated by radiation and can induce fibrosis by promoting excessive synthesis of matrix components [67]. Inhibition of TGF-β is sufficient to prevent radiation-induced differentiation of fibroblastic cells in vitro [68].

TGF-β directed gene transcription has been reported for a number of different matrix components in dermal fibroblast cultures. These include MMP2, TIMP1 and Collagen IA1 [69, 70]. Blockade of the intracellular signalling cascade mediated by TGF-β -induced Smad3 expression, through silencing of Smad3 itself, prevented transcriptional activation of the expression of these matrix genes in irradiated dermal fibroblasts [70]. Similarly, the radiation-activated transcription of type 1 collagen genes in mouse embryonic fibroblast and NIH 3T3 cells was abrogated by small-molecule inhibitors of TGF-β—receptor interaction [69]. It would appear that the TGF-β activated synthesis of extracellular matrix in response to radiation is regulated independently from the activation of the MAPK pathway regulating cell proliferation. Inhibition of TGF-β a activation of MAPK/ERK1 signalling had no influence on matrix gene transcription [69]. However, previous studies have shown that phosphorylation of ERK was required for the up-regulation of MMP2 in irradiated rat epithelial cells [71].

3.2.12 Other Considerations

3.2.12.1 Bystander Effects

Further insight into the complexity of the signalling that regulates matrix remodelling is shown by the recent study of Kühlmann et al. who show that cell-type specific production of cytokines may be critical for radiation responses involving excess matrix production. Thus, whilst TGF-β was released by irradiated fibroblasts the additional release of Interleukin 8 from epithelial cells was required for the efficient activation of matrix synthesis from fibroblasts [72].

3.2.12.2 Radiation Quality

Matrix remodelling appears largely unaffected by the quality of the radiation applied. Thus, high LET (600 MeV Fe) provoked qualitatively similar changes to the matrix composition of mammary tissue [26] and lens tissue [52]. The fibrosis-promoting effects of high LET carbon ions on collagen production was greater than comparable doses of low LET X-irradiation, and this difference was comparable to the relative biological effectiveness (RBE) seen for other biological effects such as cell killing and genetic damage [73]. TGF-β up-regulation and fibrosis was seen in the mouse kidney following alpha-irradiation [74].

However, a direct comparison between low LET (X-rays) and high LET (carbon ions) confirmed the increase in metastatic potential following X-irradiation, of HT1080 and LM8 osteosarcoma cells, but failed to show such an increase after high LET treatment. This difference does not appear to be due to differences in cell survival, nor to different degrees of up-regulation of integrin expression, as this was seen with both high and low LET treatments [75]. The high LET exposures decreased MMP2 expression, whereas the low LET did not, possibly implicating matrix lysis in the difference.

3.2.12.3 Effect of Low Doses on the Extracellular Environment

The cellular consequences of exposure to low doses of radiation are quite distinct to those seen at the therapeutic doses used for tumour control. At low doses, typically defined as those doses received from medical imaging, natural background and from the workplace, the induction of chronic diseases is more prominent than cell killing [76]. The effects of low doses on the extracellular environment are not yet fully understood. At low doses above 100 mGy mouse mammary tissue still exhibited the increase in TGF-β production seen after high radiotherapy doses [77]. This was accompanied by enhanced expression of type III collagen, suggesting that at even low doses there may be a change in the microenvironment, and possibly even a mild fibrotic response of low-dose exposed tissues.

Evidence from abnormal tissue samples taken from victims of the Chernobyl accident suggest that there is a progressive change in epithelial tissue of the bladder that is not seen in non-exposed tissues with similar disease status. Thus, in dysplastic tissue from radiation exposed individuals there is an increase in the epithelial adhesion molecule E-Cadherin in the cytoplasm, rather than in the normal membrane location. This is accompanied by increased production of TGF-β, which has already been implicated in radiation-induced changes in matrix (see above) [78]. In more advanced cancer the same increase in TGF-β and aberrant cadherin distribution were seen, along with decreased expression of fibronectin and laminin [79]. Mesenchymal cells are also influenced by low dose exposures. Human mammary gland-derived fibroblasts enter a senescence-like state when chronically irradiated at low doses. These cells activated expression of a range of MMPs, leading to rapid lysis of the Matrigel support [80].

3.3 Summary

Exposure of tumourous tissue to ionizing radiation initiates a wound-healing response involving remodelling of the extracellular microenvironment. The initial reaction involves direct damage to the matrix proteins or the secretion and activation of proteolytic enzymes that lead to local destruction of the extracellular matrix. Subsequently the wounded area may undergo complete repair, may enter a prolonged period of heightened proteolysis, or may overproduce matrix proteins leading to fibrosis. The source of matrix degrading enzymatic activity may be the tumour cells and the tumour stroma. Additional complexity is provided by proteolytic activity released from tissue macrophages, mast cells and by invading inflammatory cells. The local production of growth factors, including VEGF and TGF-β play a key role in coordinating the response. It is anticipated that the application of modern proteomic technologies will reveal hitherto unrecognised levels of complexity in these processes. Hopefully this will lead to the development of new therapeutic strategies to prevent long-term health implications of radiation exposure.

References

1. Apetoh L et al (2007) Toll-like receptor 4-dependent contribution of the immune system to anticancer chemotherapy and radiotherapy. Nat Med 13: 1050–1059
2. Quesada V, Ordóñez GR, Sánchez LM, Puente XS, López-Otín C (2009) The degradome database: mammalian proteases and diseases of proteolysis. Nucleic Acids Res 37:D239–D243
3. Mason SD, Joyce JA (2011) Proteolytic networks in cancer. Trends Cell Biol 21:228–237
4. Wagner SN et al (1995) Modulation of urokinase and urokinase receptor gene expression in human renal cell carcinoma. Am J Pathol 147:183–192
5. Lin Q, Yun Z (2010) Impact of the hypoxic tumor microenvironment on the regulation of cancer stem cell characteristics. Cancer Biol Ther 9:949–956
6. Miyazaki Y et al (2008) He effect of hypoxic microenvironment on matrix metalloproteinase expression in xenografts of human oral squamous cell carcinoma. Int J Oncol 32:145–151

7. Erler JT et al (2009) Hypoxia-induced lysyl oxidase is a critical mediator of bone marrow cell recruitment to form the premetastatic niche. Cancer Cell 15:35–44
8. Kakkad SM et al (2010) Hypoxic tumor microenvironments reduce collagen I fiber density. Neoplasia 8:608–617
9. Indovina P et al (2006) Hypoxia and ionizing radiation: changes in adhesive properties and cell adhesion molecule expression in MG-63 three-dimensional tumor spheroids. Cell Commun Adhes 13:185–198
10. Myllyharju J, Schipani E (2010) Extracellular matrix genes as hypoxia-inducible targets. Cell Tissue Res 339:19–29
11. Thiely JP (2002) Epithelial-mesenchymal transitions in tumor progression. Nat Rev Cancer 2:442–454
12. Jean C et al (2011) Influence of stress on extracellular matrix and integrin biology. Oncogene 30:2697–2706
13. Tibbs MK (1997) Wound healing following radiation therapy: a review. Radiother Oncol 42:99–106
14. Mueller CK, Schultze-Mosgau S (2009) Radiation-induced microenvironments–the molecular basis for free flap complications in the pre-irradiated field? Radiother Oncol 93:581–585
15. Kruegler WWO et al (1978) Fibroblast implantation enhances wound healing as indicated by breaking strength determinations. Otolaryngology 86:804–811
16. Penney DP, Rosenkrans WA (1984) Cell-cell matrix interactions in induced lung injury. I. The effects of X-irradiation on basal laminar proteoglycans. Radiat Res 99:410–419
17. Carnevali S et al (2003) Gamma radiation inhibits fibroblast-mediated collagen gel retraction. Tissue Cell 35:459–469
18. Wang J, Zheng H, Hauer-Jensen M (2001) Influence of short-term octreotide administration on chronic tissue injury, transforming growth factor β (TGF-β) overexpression, and collagen accumulation in irradiated rat intestine. J Pharmacol Exp Ther 297:35–42
19. Rodemann HP, Bamberg M (1995) Cellular basis of radiation-induced fibrosis. Radiother Oncol 35:83–90
20. Riedel F et al (2005) Immunohistochemical analysis of radiation-induced non-healing dermal wounds of the head and neck. In Vivo 19:343–350
21. Kim K, McBride WH (2010) Modifying radiation damage. Curr Drug Targets 11:1352–1365
22. Açil Y et al (2007) Proof of direct radiogenic destruction of collagen in vitro. Strahlenther Onko 183:374–379
23. Springer IN et al (2005) Radiation caries–radiogenic destruction of dental collagen. Oral Oncol 41:723–728
24. Hübner W et al (2005) The influence of X-ray radiation on the mineral/organic matrix interaction of bone tissue: an FT-IR microscopic investigation. Int J Artif Organs 28:66–73
25. Gouk SS et al (2008) Alterations of human acellular tissue matrix by gamma irradiation: histology, biomechanical property, stability, in vitro cell repopulation, and remodeling. J Biomed Mater Res B Appl Biomater 84:205–217
26. Ehrhart EJ, Gillette EL, Barcellos-Hoff MH (1996) Immunohistochemical evidence of rapid extracellular matrix remodeling after iron-particle irradiation of mouse mammary gland. Radiat Res 145:157–162
27. Miller GG, Kenning JM, Dawson DT (1988) Radiation-induced changes in collagen isotypes I, III, and IV in the lung of LAF1 mouse: effects of time, dose, and WR-2721. Radiat Res 115:515–532
28. Rosenkrans W-A, Penney DP (1987) Cell-cell matrix interactions in induced lung injury. IV. Quantitative alterations in pulmonary fibronectin and laminin following X irradiation. Radiat Res 109:127–142
29. Giannopoulou E et al (2001) X-rays modulate extracellular matrix in vivo. Int J Cancer 94:690–698
30. Vozenin-Brotons MC et al (2004) Gene expression profile in human late radiation enteritis obtained by high-density cDNA array hybridization. Radiat Res 161:299–311
31. Bourgier C et al (2005) Inhibition of Rho kinase modulates radiation induced fibrogenic phenotype in intestinal smooth muscle cells through alteration of the cytoskeleton and connective tissue growth factor expression. Gut 54:336–343
32. Haydont V et al (2007) Pravastatin inhibits the Rho/CCN2/extracellular matrix cascade in human fibrosis explants and improves radiation-induced intestinal fibrosis in rats. Clin Cancer Res 13:5331–5340
33. Williams JP et al (2004) Effect of administration of lovastatin on the development of late pulmonary effects after whole-lung irradiation in a murine model. Radiat Res 161:560–567
34. Angenete E et al (2009) Preoperative radiotherapy and extracellular matrix remodeling in rectal mucosa and tumour matrix metalloproteinases and plasminogen components. Acta Oncol 48:1144–1151
35. Zhao W et al (2001) Redox modulation of the profibrogenic mediator plasminogen activator inhibitor-1 following ionizing radiation. Cancer Res 61:5537–5543
36. Zhao W et al (2000) Irradiation of rat tubule epithelial cells alters the expression of gene products associated with the synthesis and degradation of extracellular matrix. Int J Radiat Biol 76:391–402
37. Green N et al (1969) Radiation-induced delayed union of fractures. Radiology 93:635–641
38. Arnold M, Kummermehr J, Trott K-R (1995) Radiation-induced impairment of osseous healing: quantitative studies using a standard drilling defect in rat femur. Radiat Res 143:77–84
39. Matsumura S et al (1998) Changes in phenotypic expression of osteoblasts after X irradiation. Radiat Res 149:463–471
40. Dudziak ME et al (2000) The effects of ionizing radiation on osteoblast-like cells in vitro. Plast Reconstr Surg 106:1049–1061

41. Narayan K, Cliff WJ (1982) Morphology of irradiated microvasculature: a combined in vivo and electron-microscopic study. Am J Pathol 106:47–62
42. Neta R, Oppenheim JJ, Douches SD (1988) Interdependence of the radioprotective effects of human recombinant interleukin 1 alpha, tumor necrosis factor alpha, granulocyte colony-stimulating factor, and murine recombinant granulocyte-macrophage colony-stimulating factor. J Immunol 140:108–111
43. Zhou Y et al (2010) Modulation of matrix metalloproteinase-9 and tissue inhibitor of metalloproteinase-1 in RAW264.7 cells by irradiation. Mol Med Rep 3:809–813
44. Müller K, Meineke V (2011) Radiation-induced mast cell mediators differentially modulate chemokine release from dermal fibroblasts. J Dermatol Sci 61:199–205
45. Steiner HH et al (2004) Autocrine pathways of the vascular endothelial growth factor (VEGF) in glioblastoma multiforme: clinical relevance of radiation-induced increase of VEGF levels. J Neurooncol 66:129–138
46. Polytarchou C et al (2004) X-rays affect the expression of genes involved in angiogenesis. Anticancer Res 24:2941–2945
47. Brooks P, Roth JM, Lymberis SC, DeWyngaert K, Broek D, Formenti SC (2002) Ionizing radiation modulates the exposure of the HUIV26 cryptic epitope within collagen type IV during angiogenesis. Int J Radiat Oncol Biol Phys 54:1194–1201
48. Carmeliet P, Collen D (1998) Development and disease in proteinase-deficient mice: role of the plasminogen, matrix metalloproteinase and coagulation system. Thromb Res 91:255–258
49. Takeuchi H et al (2012) A mechanism for abnormal angiogenesis in human radiation proctitis: analysis of expression profile for angiogenic factors. J Gastroenterol 47:56–64
50. Yang K et al (2007) Matrix-metallo-proteinases and their tissue inhibitors in radiation-induced lung injury. Int J Radiat Biol 83:665–676
51. Lee WH et al (2012) Irradiation alters MMP-2/TIMP-2 system and collagen type IV degradation in brain. Int J Radiat Oncol Biol Phys 82:1559–1566
52. Chang PY et al (2007) Particle radiation alters expression of matrix metalloproteases resulting in ECM remodeling in human lens cells. Radiat Environ Biophys 46:187–194
53. Abderrahmani R et al (2009) Effects of pharmacological inhibition and genetic deficiency of plasminogen activator inhibitor-1 in radiation-induced intestinal injury. Int J Radiat Oncol Biol Phys 74:941–948
54. Gogineni VR et al (2009) RNAi-mediated downregulation of radiation-induced MMP-9 leads to apoptosis via activation of ERK and Akt in IOMM-Lee cells. Int J Oncol 34:209–218
55. Cheng JC et al (2006) Radiation-enhanced hepatocellular carcinoma cell invasion with MMP-9 expression through PI3K/Akt/NF-kappaB signal transduction pathway. Oncogene 25:7009–7018
56. Park CM et al (2006) Ionizing radiation enhances matrix metalloproteinase-2 secretion and invasion of glioma cells through Src/epidermal growth factor receptor-mediated p38/Akt and phosphatidylinositol 3-kinase/Akt signaling pathways. Cancer Res 66:8511–8519
57. Schweigerer L et al (2005) Sublethal irradiation promotes invasiveness of neuroblastoma cells. Biochem Biophys Res Commun 330:982–988
58. Park CC et al (2003) Ionizing radiation induces heritable disruption of epithelial cell interactions. Proc Natl Acad Sci U S A 100:10728–10733
59. Tsukamoto H et al (2007) Irradiation-induced epithelial-mesenchymal transition (EMT) related to invasive potential in endometrial carcinoma cells. Gynecol Oncol 107:500–504
60. Wang M et al (2012) Heavy ions can enhance TGFβ mediated epithelial to mesenchymal transition. J Radiat Res 53:51–57
61. Cordes N et al (2006) Beta1-integrin-mediated signaling essentially contributes to cell survival after radiation-induced genotoxic injury. Oncogene 25:1378–1390
62. Fuks Z et al (1992) Effects of extracellular matrix on the response of endothelial cells to radiation in vitro. Eur J Cancer 28A:725–731
63. Simon EL et al (2005) High dose fractionated ionizing radiation inhibits prostate cancer cell adhesion and beta(1) integrin expression. Prostate 64:83–91
64. Park CC et al (2008) Beta1 integrin inhibition dramatically enhances radiotherapy efficacy in human breast cancer xenografts. Cancer Res 68:4398–4405
65. Monferran S et al (2008) Alphavbeta3 and alphavbeta5 integrins control glioma cell response to ionising radiation through ILK and RhoB. Int J Cancer 123:357–364
66. Nam JM et al (2010) Breast cancer cells in three-dimensional culture display an enhanced radioresponse after coordinate targeting of integrin alpha5beta1 and fibronectin. Cancer Res 70:5238–5248
67. Ise K et al (2004) Transforming growth factor-beta signaling is enhanced following mitomycin-C treatment of islet xenograft. Transplant Proc 36:1183–1185
68. Hakenjos L, Bamberg M, Rodemann HP (2000) TGF-beta1-mediated alterations of rat lung fibroblast differentiation resulting in the radiation-induced fibrotic phenotype. Int J Radiat Biol 76:503–509
69. Yano H et al (2010) Smad, but not MAPK, pathway mediates the expression of type I collagen in radiation induced fibrosis. Biochem Biophys Res Commun 418:457–463
70. Lee JW et al (2010) Regulators and mediators of radiation-induced fibrosis: gene expression profiles and a rationale for Smad3 inhibition. Otolaryngol Head Neck Surg 143:525–530
71. Zhao W, Goswami PC, Robbins ME (2004) Radiation-induced up-regulation of Mmp2 involves increased mRNA stability, redox modulation, and MAPK activation. Radiat Res 161:418–429

72. Kuhlmann UC et al (2009) Radiation-induced matrix production of lung fibroblasts is regulated by interleukin-8. Int J Radiat Biol 85:138–143
73. Fournier C et al (2001) Changes of fibrosis-related parameters after high- and low-LET irradiation of fibroblasts. Int J Radiat Biol 77:713–722
74. Jaggi JS et al (2005) Renal tubulointerstitial changes after internal irradiation with alpha-particle-emitting actinium daughters. J Am Soc Nephrol 16:2677–2689
75. Ogata T et al (2005) Particle irradiation suppresses metastatic potential of cancer cells. Cancer Res 65:113–120
76. Mullenders L et al (2009) Assessing cancer risks of low-dose radiation. Nat Rev Cancer 9:596–604
77. Ehrhart EJ et al (1997) Latent transforming growth factor beta1 activation in situ: quantitative and functional evidence after low-dose gamma-irradiation. FASEB J 11:991–1002
78. Romanenko A et al (2006) Aberrant expression of E-cadherin and beta-catenin in association with transforming growth factor-beta1 in urinary bladder lesions in humans after the Chernobyl accident. Cancer Sci 97:45–50
79. Romanenko A et al (2006) Extracellular matrix alterations in conventional renal cell carcinomas by tissue microarray profiling influenced by the persistent, long-term, low-dose ionizing radiation exposure in humans. Virchows Arch 448:584–590
80. Tsai KK et al (2005) Cellular mechanisms for low-dose ionizing radiation-induced perturbation of the breast tissue microenvironment. Cancer Res 65:6734–6744

Serum and Plasma Proteomics and Its Possible Use as Detector and Predictor of Radiation Diseases

Olivier Guipaud

Abstract

All tissues can be damaged by ionizing radiation. Early biomarkers of radiation injury are critical for triage, treatment and follow-up of large numbers of people exposed to ionizing radiation after terrorist attacks or radiological accident, and for prediction of normal tissue toxicity before, during and after a treatment by radiotherapy. The comparative proteomic approach is a promising and powerful tool for the discovery of new radiation biomarkers. In association with multivariate statistics, proteomics enables measurement of the level of hundreds or thousands of proteins at the same time and identifies set of proteins that can discriminate between different groups of individuals. Human serum and plasma are the preferred samples for the study of normal and disease-associated proteins. Extreme complexity, extensive dynamic range, genetic and physiological variations, protein modifications and incompleteness of sampling by two-dimensional electrophoresis and mass spectrometry represent key challenges to reproducible, high-resolution, and high-throughput analyses of serum and plasma proteomes. The future of radiation research will possibly lie in molecular networks that link genome, transcriptome, proteome and metabolome variations to radiation pathophysiology and serve as sensors of radiation disease. This chapter reviews recent advances in proteome analysis of serum and plasma as well as its applications to radiation biology and radiation biomarker discovery for both radiation exposure and radiation tissue toxicity.

Keywords

Bioindicator • Biomarker • Diagnostic • Ionizing radiation • Irradiation • Normal tissue toxicity • Omics • Prognostic • Plasma • Proteome • Proteomics • Radiation exposure • Radiotherapy • Serum

---

O. Guipaud (✉)
Institute for Radiological Protection and Nuclear Safety
(IRSN), PRP-HOM, SRBE, LRTE, BP17,
Fontenay-aux-Roses cedex 92262, France
e-mail: olivier.guipaud@irsn.fr

Abbreviations

1D-SDS-PAGE	One-dimensional sodium dodecyl sulfate polyacrylamide gel electrophoresis
2D-DIGE	Two-dimensional in-gel differential gel electrophoresis
2D-GE	Two-dimensional gel electrophoresis
Ab	Antibody
AEC	Anion exchange chromatography
ApoA-1	Apolipoprotein A-1
APP	Acute phase protein
BALF	Bronchoalveolar fluid
Da	Dalton
ELISA	Enzyme-linked immunosorbent assay
FFE	Free flow electrophoresis
Flt3-L	Flt3-ligand
G-CSF	Granulocyte colony-stimulating factor
Gy	Gray
HC	Hierarchical clustering
HMW	High molecular weight
HPLC	High-performance liquid chromatography
HUPO	Human proteome organization
IEF	Isoelectric focusing
IFN-γ	Interferon gamma
IL	Interleukin
IL-1ra	Interleukin 1 receptor antagonist
IMRT	Intensity-modulated radiation therapy
IP10	Interferon gamma-induced protein 10
IR	Irradiation
iTRAQ	Isobaric tags for relative and absolute quantitation
KEGG	Kyoto encyclopedia of genes and genomes
KL-6	Krebs von den Lungen-6
LC	Liquid chromatography
LMW	Low molecular weight
LTGF-β	Latent transforming growth factor beta
MCP-1	monocyte chemotactic protein-1
METREPOL	Medical treatment protocols for radiation accident victims
MHC I H2-Q10 α chain	Major histocompatibility complex I histocompatibility 2 Q region locus 10 alpha chain
MRM	Multiple reaction monitoring
MS	Mass spectrometry
MS/MS	Tandem mass spectrometry
NSCLC	Non-small cell lung cancer
PCA	Principal component analysis
PF2D	Chromatofocusing-reverse phase-liquid chromatography
pI	Isoelectric point
PLS-DA	Partial least square discriminant analysis
PPP	Plasma proteome project
Pzp	Pregnancy zone protein
Q-TOF	Quadrupole time-of-flight
RP-LC	Reverse phase liquid chromatography
RT	Radiation therapy
SCX	Strong cation exchange
SDS-PAGE	Sodium dodecyl sulfate polyacrylamide gel electrophoresis
SEC	Size-exclusion chromatography
SELDI-TOF	Surface-enhanced laser desorption/ionization time-of-flight
SRM	Selected reaction monitoring
TGF-β	Transforming growth factor beta
TNF-α	Tumor necrosis factor alpha
TOF	Time-of-flight
α2M	α-2-macroglobulin

4.1 Introduction

Ionizing radiation causes injury of normal tissue by a dynamic and developing process which involves cell killing, altered cell-to-cell communications, inflammatory responses, compensatory tissue hypertrophy of remaining normal tissue, and tissue repair processes [1, 2]. Changes in cytokine levels can be detected over time [3, 4]. Tissues with fast renewal present effects which are mainly due to the disappearance of cells. Tissues with slow renewal, or which do not proliferate, present effects related to inflammation and devascularization. The process which leads to the toxicity is sequential and progressive in most tissues, but the rate of progression varies between tissues, and differs from patient to patient. The genetic characteristics of the exposed organism can also influence the exact nature of the radiation response [5]. The toxicity of some critical organs is discussed in the next section.

Damage induced by ionizing radiation is usually classified according to the time of appearance of the symptoms: acute effects, late consequential effect or late effects [2]. Acute, or early, effects are observed rapidly after exposure, in the weeks which follow the end of radiotherapy, or even during the treatment. Consequential late effects appear later and are caused by persistent acute damage the grade or duration of which influences the severity [6]. Late effects appear months or years after exposure. The cellular and molecular processes which govern the damage, and its repair, are complex and lead to many events. For some organs which develop late damage, the early symptoms cannot be visible. Trauma or surgery can then trigger an acute episode in the organ, which until then was apparently working normally.

Given the growing threat of terrorist attacks designed to cause massive human losses at unpredictable locations and the risk of nuclear power plant catastrophes, as illustrated by the recent Fukushima Daiichi nuclear disaster, the guidelines for radiation or nuclear accident management should be reformulated. In situations caused by terrorist attacks or nuclear power plant accidents, the number of victims will depend on the configuration of the event, but could range in the case of a dirty bomb or the malevolent dispersion of an orphan source from several tens or hundreds of victims to thousands or hundreds of thousands of victims in the case of a nuclear device. Victims can vary markedly in terms of the severity and type of exposure. This extreme variation makes it very difficult to harmonize and standardize the relevant diagnosis, prognosis and treatment methods. The European approach for medical preparedness has designed optimal methodology for triage of victims [7]. The first 48 h after a radiological accident involving masses of people are crucial. In that time, the accident victims should be processed by an emergency triage system where the patients are scored on the basis of both clinical and biological criteria. A primary objective in the first 48 h is to identify individuals who were not irradiated. After the initial 48 h, scoring of the patient is re-evaluated on the basis of METREPOL (medical treatment protocols for radiation accident victims) [8, 9]. Unfortunately it is currently not possible to know, during the first 48 h, the individual physical and biological dosimetry, which only becomes accessible after 48 h, at which point it forms the basis of further medical decision-making.

On the other hand, the use of medical therapeutic radiation for most types of cancer has improved dramatically over the last decades. Normal tissues surrounding the tumor are irradiated and may develop symptomatic injury which is known to develop immediately and to progress long after exposure to radiation [2]. Total and daily (fractionation) irradiation doses, type of ionizing radiation, field size, time of treatment, use of surgery or chemotherapy influence the extent of normal tissue injury. The tolerance of these normal tissues determines the dose prescription for the treatment of most cancers.

Biomarkers to monitor a potentially exposed population after a radiological accident or to assess radiation toxicity before, in the course of, or after a treatment by radiotherapy should be developed and would be extremely valuable for estimating the risks associated with radiation exposure. Rapid detection of radiation injury by high-throughput automated equipment or a point of care assay are needed for early diagnosis to prevent loss of organ function or mortality in order to screen large numbers of people rapidly and accurately. Biomarkers would allow medical personnel to make critical triage decisions. Since the METREPOL approach is based on clinical symptoms, there is a need to develop other markers to reinforce existing protocols, particularly when the date of exposure is not known or when the accident is disclosed beyond the first 48 h. These biomarkers should be easy to collect using non- or semi-invasive methods, as well as easy and fast to quantify. They should help to determine quickly if a person has been exposed or not, and whether the exposure was limited to a small area or concerned the whole body. Ideally, they should provide information on the damaged organ or tissue, and help to predict the extent of the upcoming damage. For example, following a radioactive source accident, the victim can be severely exposed locally but it is currently impossible to predict the outcome of the cutaneous lesion (i.e. necrosis or not). In such cases it is critical to develop biomarkers which predict the severity of the lesion.

As organisms respond to irradiation by altering the expression and/or the post-translational modifications of some proteins in cells, tissues and organic fluids, as serum, plasma or urine, it is conceivable that protein expression profiling can be used to define protein or, much better, a set of protein expression changes that differentiates between irradiated and non-irradiated individuals, or that differentiates early between detrimental and harmless upcoming injuries in the case of restricted exposures, because of a treatment by radiotherapy, for instance. Proteins are easily obtained using non-invasive (urine) or semi-invasive (blood) collection methods. Using immunodetection techniques, their quantifications in biological fluids (urine, serum), cells (circulating lymphocytes) or tissue (biopsy material) are fast and reliable. In addition, automation is not problematic for high-throughput analysis of thousands of people. Lastly, today proteomic and multivariate statistical analysis methods provide a unique tool to look for a set of proteins that can discriminate between different health statuses. Recent advances in proteome analysis of serum and plasma as well as its applications to radiation biology and radiation biomarker discovery are reviewed in the next sections of this chapter.

4.2 Normal Tissue Radiation Toxicity

All tissues are damaged by ionizing radiation [10]. The damage in normal tissues results from the initial deposition of energy within the tissue. The pathological processes of the radiation injury begin at once after the irradiation, but the clinical and histological signs cannot become visible before weeks, months, or even years after an accidental exposure or after a therapeutic treatment [2, 10].

The acute effects are mainly observed in tissues with fast renewal, for example epithelia of the skin surface or the digestive tract. The symptoms develop when functional cells are lost, because of the turnover of the normal tissue, and are not replaced because of the damage sustained by the stem cell compartment. In tissues as the skin or the bowel, there are compensatory phenomena involving the proliferation of stem cells, which are more tolerant to ionizing radiation than the other cells and replace the cells of the tissue thereby allowing its repair. The symptoms remain during this phase, which may continue throughout radiotherapy treatment. Cells die generally during mitosis, in the course of one or more of the first divisions which follow irradiation, because of damaged chromosomes which are not or badly repaired (mitotic death). Other cellular types, for example lymphocytes, can die by apoptosis, a fast death in particular tissue locations. Cells can also leave the reproductive pool by differentiation [11]. This cellular senescence is a particularly important response of fibroblasts, which can lead to excess deposition of collagen and subsequently to fibrosis. Early effects, such as erythema of the skin, probably involve mechanisms other than cellular death. Irradiation activates various cell signaling pathways [12], leading to the expression and activation of proinflammatory and profibrotic cytokines [3, 13–15], vascular damage [16, 17], and activation of coagulation [18]. These changes can then lead to edema, inflammatory responses, and initiation of a healing-like process: waves of cytokines are produced in an effort to cure the damage [3, 19].

Late effects develop months to years after an exposure and are mainly observed in patients undergoing radiotherapy. The symptoms can be moderate to severe, limited or progressive, and can develop gradually or suddenly. The late effects tend to develop in tissues containing cells which are slowly renewed (subcutaneous tissue, fatty tissue, muscle, brain, kidney, liver), or sites of low renewal inside tissues which contain cells which quickly proliferate (wall of the intestine). Usual tissue lesions are fibrosis, necrosis, atrophy, and vascular damage. Late effects develop through complex processes which interact together [20] and which are not currently fully understood, in particular regarding the role of cell death. Cells are part of a community in which the members depend on each other and contribute individually to the functioning of the tissue and the whole body. The production and release of cytokines and growth factors are an integral part of the response to irradiation and lead to an adaptive response and to the infiltration of cells related to the mechanism of healing. This process is perpetuated by cell death and consequent loss of cells, by the deregulation of interactions between surviving cells, and by the hypoxia induced by the vascular damage.

The most common pathological changes in patients treated by radiotherapy take place in epithelia (parenchyma), stroma (mesenchyma) and the vasculature [21]. Epithelial changes include mainly atrophy, necrosis and ulceration. Necrosis and ulceration generally cohabit and precede atrophy. It is unusual to observe necrosis without ulceration, except in the white matter of the central nervous system. Atrophy is characterized by a progressive decrease in the number and volume of epithelial cells. Any tissue which contains epithelial cells can present atrophic changes (skin, subcutaneous tissue, gastrointestinal system, genitourinary organs, respiratory tracts, breast, and salivary glands). Stromal changes include mainly fibrosis. Fibrosis is a common sequela of both cancer treatment by radiotherapy and accidental irradiation and has been described in many tissues including skin, lung, heart and liver. The underlying mechanisms of the radiation-induced fibrosis have yet to be elucidated in detail [22]. The most frequent sites of radiation-induced fibrosis are head and neck regions, breast, genitourinary and gastrointestinal tissues. Some organs, such as the central nervous system and the lens of the eye, rarely develop fibrosis. This is also the case for bone marrow, which responds with serous atrophy and fatty replacement, but rarely fibrosis. Finally, irradiation of the liver and bone induces necrosis rather than fibrosis. Vascular changes contribute to secondary damage of the organ because of decreased perfusion which induces ischemia, necrosis, ulceration, or fibrosis. Endothelial cells are the most radiosensitive cells of the vasculature, but the size of blood vessels determine their susceptibility to the radiation-induced damage. The most common morphological changes are damage to the endothelium, intimal hyperplasia, vascular wall fibrosis and microthrombus formation [16].

In conclusion, radiation-induced morphologic alterations can be seen in nearly every organ, vary from organ to organ, and can be broadly categorized as injury of epithelial, stromal and vascular tissues. Diagnosis and prognosis of tissue and organ toxicities remain a challenge in the field of radiation research, in particular because all organs are likely to respond differently to ionizing radiation, and since ionizing radiation mostly affects several organs and tissues at the same time.

4.3 Overview of Serum and Plasma Proteome Research and Application to Human Disease Detection

Omics dominates the list of new techniques available to the researcher studying almost any biological system. Rapid developments in technology platforms coupled with high-throughput techniques allow researchers to measure many analytes from many samples simultaneously, generating vast amounts of data. The fields of bioinformatics and multivariate statistics, coupled with increasing computational power, give the researcher solutions to problems associated with data-mining as well as opportunities for research and development of novel solutions.

Proteomic methodologies offer exciting opportunities for discovering diagnostic and prognostic markers of a number of important diseases. Single disease biomarkers have the shortcomings of insufficient sensitivity and specificity, which may be enhanced by the measurement of a combination of biomarkers, discovered using proteomics. Proteomics aims to describe and characterize all proteins expressed in a biological system. In theory, the comparative proteomic approach enables measurement of the level of hundreds or thousands of proteins at the same time, without any a priori information about the identity of proteins. In addition, the results can be analyzed using statistical multivariate tools, to determine the smallest number of variables (set of proteins) required for the discrimination of different groups of individuals.

Proteomics technologies are broadly characterized as tools designed to examine the expression of proteins in a given set of samples. Types of information derived from proteomics technologies include protein identity, quantity, interactions, structure, and post-translational modification. The nature of the proteome presents a series of challenges that limit the ability of any single technology to assay and characterize it completely. The main problem is the heterogeneity of proteins. Proteins can be large or small, globular or compact, hydrophobic or hydrophilic, basic or acidic. Post-translational modifications such as glycosylation, lipidation or phosphorylation can have dramatic effects on how the proteins behave in biochemical assays. A second challenge of the proteome is the extraordinarily wide dynamic range of protein expression, which is estimated to be more than 10^{12} [23].

Human body fluids are all precious resources for discovery of human disease biomarkers, and in particular for radiation biomarker discovery. They include serum, plasma, urine, lymph, cerebrospinal fluid, saliva, bronchoalveolar lavage fluid, synovial fluid, nipple aspirate fluid, tear fluid and amniotic fluid [24]. Body fluids contain a large number of proteins possibly modified in a variety of forms. The variability of body fluid proteomes is huge due to their complexity and must be considered in the design of the experiment. The urine proteome has been investigated for radiation biomarker discovery [25] and constitutes the scope of the Sect. 3.5. Urine has the great advantage over serum and plasma of being collected easily in a non-invasive manner. However, blood samples are mostly considered homogeneous when compared with saliva or urine, the compositions of both of which are rather dependent on fluid flow rates.

Serum and plasma proteins originate from a variety of tissues and blood cells as a result of secretion or leakage, but almost all the plasma proteins are synthesized in liver except gamma globulins. Serum and plasma protein levels reflect human physiological or pathological states and can be used for disease diagnosis and prognosis [26, 27]. Plasma contains a huge number of proteins differing by a dynamic range of at least 12 orders of magnitude [26]. Many of these proteins are post-translationally modified (mostly glycosylated), adding to the diversity and complexity of serum and plasma proteins. Partially degraded proteins or protein fragments add variability to the complexity of the serum and the plasma proteomes [27]. There are an increasing number of publications related to serum and plasma proteomics and biomarkers (Fig. 4.1), showing that the scientific community extensively trusts in proteome research for biomarker discovery studies.

Serum and plasma protein compositions differ greatly [28]. As a consequence of the coagulation process in the preparation of the serum, fibrinogen and associated proteins such as high-molecular-weight von Willebrand factor are removed from plasma and variable addition of cellular secretion products can occur. Whether to use serum or plasma for proteome analysis is a question still to be solved. The Human Proteome Organization (HUPO) Plasma Proteome Project (PPP) Specimens Committee concluded that plasma was preferable to serum due to less degradation ex vivo [29–31]. However, sera are

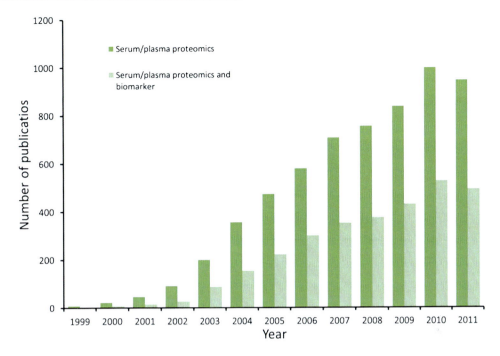

Fig. 4.1 Increasing number of publications in the field of serum/plasma proteomics and biomarker discovery. The histogram illustrates the number of hits for serum/plasma proteomics (*dark green bars*) and for serum/plasma proteomics and biomarkers (*light green bars*) in PubMed (http://www.ncbi.nlm.nih.gov/pubmed) using the respective combinations of keywords : (1) "(serum AND proteome) OR (serum AND proteomics) OR (serum AND proteomic) OR (plasma AND proteome) OR (plasma AND proteomics) OR (plasma AND proteomic)"; (2) "((serum AND proteome) OR (serum AND proteomics) OR (serum AND proteomic) OR (plasma AND proteome) OR (plasma AND proteomics) OR (plasma AND proteomic)) AND (marker OR markers OR biomarker OR biomarkers OR bioindicator OR bioindicators)". The number of publications has regularly increased during the last decade and perhaps started to stabilize in 2011, reaching a maximum of 900–1,000 publications a year for serum/plasma proteomics, and 400–500 publications a year for serum/plasma proteomics and biomarkers

frequently used as archived specimens, so there is still debate on whether serum or plasma should be used for biomarker discovery using proteome analysis.

A major challenge of proteomic analysis of serum and plasma is the low abundance proteins. The main serum and plasma proteins are albumin (more than 50% of the whole content), present at high concentration, immunoglobulins, fibrinogen (only in plasma), transferrin, haptoglobin, alpha-antitrypsin and lipoproteins. Fibrinogen is easily removable by clotting. Several methodologies have been reported to remove high-abundance proteins prior to proteome analysis [32–35]. Unfortunately, these methods lead to the disappearance of many proteins and peptides with potential clinical relevance. The serum and plasma proteome is a dynamic entity influenced for instance by sleep, sport training, diet, and pregnancy. It is therefore of importance to set up the best preanalytical conditions when mining the serum/plasma proteome and to standardize the conditions of collection to enable comparative studies.

Great progress has been achieved in characterization of the serum/plasma proteome [36, 37]. Recently, Farrah et al. used advanced computational methods, developed for the analysis and integration of very large and diverse data sets generated by tandem mass spectrometry (MS/MS) measurements of tryptic peptides, to compile a high-confidence human plasma proteome reference set with well over twice the identified proteins of previous high-confidence sets [38]. They identified 20,433 distinct peptides, from which they inferred a highly non-redundant set of 1,929 protein sequences. They have made this resource available via PeptideAtlas, a large, multiorganism, freely accessible compilation of peptides identified in MS/MS experiments conducted by laboratories around the world (www.peptideatlas.org) [39].

The low-molecular-weight (LMW) fraction of the serum/plasma proteome is also an invaluable source of information in the context of identifying blood-based biomarkers of disease. Peptidomics aims to analyze the peptidome and to discover LMW biomarkers of human diseases [40]. The LMW fraction is obtained after the removal of high-molecular-weight proteins using physical methods such as ultrafiltration, gel chromatography and precipitation [41]. The LMW fraction is made up of several key proteins such as cytokines, chemokines, peptide hormones, as well as fragments of larger proteins. Many components of the LMW fraction of serum/plasma are bound to specific carrier proteins, albumin and non-albumin carriers, which may yield essential information [42, 43].

Many proteins are glycosylated in the serum and the plasma. Changes in the extent of the glycosylation and the glycan structures of proteins have been linked to cancer and other disease states, highlighting the clinical importance of these modifications as an indicator of pathology. The whole glycosylated status of tissues, cells or body fluids can be studied using emerging glycomic and glycoproteomic tools [44], whose great potential for deepening understanding of diseases and for the discovery of new biomarkers is now clearly established [45–48]. This contribution is essentially due to technical advances made with mass spectrometry-based investigations [49, 50].

A general scheme for the finding of serum or plasma biomarkers is depicted in Fig. 4.2. The ultimate step will be to design a clinical test for the diagnosis or the prognosis of the disease. The first step is the identification of a set of proteins from serum or plasma that discriminate classes of samples from each other. Proteomic tools and statistical and computational methods are at the core of this step since they allow collection and assembly of the few variables, among thousands, that are sufficient to discriminate several classes of samples statistically. All proteomic methodologies are based on the same principle, which is the separation of proteins or peptides prior to analysis by mass spectrometry (MS) or antibody array. In the classic two-dimensional electrophoresis method (gel-based method), significantly improved by the use of fluorescent dyes (two-dimensional in-gel differential gel electrophoresis, 2D-DIGE), the proteins are separated in polyacrylamide gels according to their charge (isoelectric focusing, IEF) and molecular mass and shape (SDS-PAGE). Protein variants are then recognized by software-based image and statistical analysis, and identified by MS analysis. In higher throughput methods (gel-free methods), complex protein or peptide mixtures are separated by multidimensional liquid chromatography (LC) (shotgun proteomics) or by affinity surfaces or beads (ProteinChip®, CLINPROT™) prior to direct analysis by tandem mass spectrometry (LC-MS/MS) or by time-of-flight mass spectrometry (surface-enhanced laser desorption/ionization time-of-flight MS, SELDI-TOF-MS). In gel-free methods, because of upper limits on mass detection of mass spectrometers, proteins need to be digested into peptides and the peptides are then separated, identified and quantified from this complex enzymatic digest. The problem in digesting proteins first and then analyzing the peptide cleavage fragments by mass spectrometry is that huge numbers of peptides are generated and this prevents direct mass spectrometry analyses. The liquid chromatography approach to proteomics allows fractionation of peptide mixtures to enable and maximize identification and quantification of the component peptides by mass spectrometry. Finally, antibody-based proteomics can also be used to discover new serum/plasma biomarkers of diseases, but in this case the assay will greatly depend on available antibodies. The next step is the confirmation of the changes found by other methods, usually using antibody-based assays like ELISA, Luminex™ system or Western-blot. The identified candidate markers

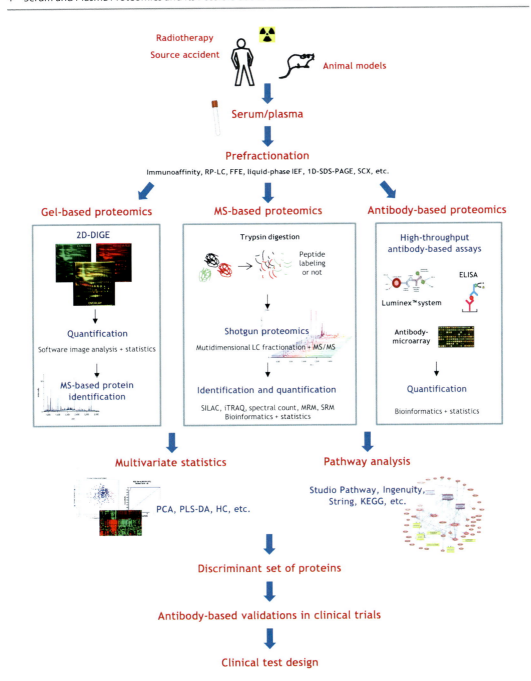

Fig. 4.2 Application of proteomics technologies to expression profiling of radiation exposure and radiation tissue toxicity using serum and plasma samples. Protein profiling is currently achieved using three main methodologies: gel-based proteomics, MS-based (or gel-free) proteomics, and antibody-based proteomics. Prior to proteomics analysis, protein samples are preferentially fractionated in order to reduce their complexity, thus increasing the chance of the identification and quantification of less abundant proteins. In turn, multifractionation of serum/plasma proteins leads to a huge increase in the number of fractions to analyze, which is incompatible with current times of analysis using proteomics tools, in particular if huge numbers of samples have to be analyzed (clinical studies). Once proteins are identified, the use of multivariate statistics is the only solution to improving sensitivity and specificity of radiation exposure detection and tissue toxicity prediction. Software based on text-mining algorithms can be used to explore molecular pathways and to propose new target proteins related to identified and quantified proteins. Candidates have to be validated, mostly by using antibody-based assays prior to devising a clinical test likely based on antibody detection

can alternatively be verified using quantitative multiple reaction monitoring (MRM) MS which has improved both the specificity and sensitivity of MS-based assays. Once the protein variants have been validated, the last step will be the design of a multiparametric test for clinical application. Such tests will be based on antibody recognition and performed using the Luminex™ system or antibody microarrays, which are described below. The use of SELDI-TOF technology for fast protein profiling of biological samples also remains a promising technique with clinical applications [51, 52].

Multiplexed proteomics platforms, which often consist of prefractionation, multidimensional separations, and MS/MS techniques, are required to achieve a more widespread analysis. Prefractionation of serum/plasma proteins can be performed by using a number of separation techniques, including reverse phase-LC (RP-LC), strong cation exchange (SCX) chromatography, size-exclusion chromatography (SEC), anion exchange chromatography (AEC), anion displacement liquid chromatofocusing chromatography, 1D-SDS-PAGE (slices of gel bands), membrane-based electrophoresis (Gradiflow™ electrophoresis technology), free flow electrophoresis (FFE), liquid-phase IEF, and chromatofocusing-RP-LC (PF2D) [53–60]. In order to enhance the resolving power, these techniques can be multiplexed, for instance by combining SEC with AEC or solution IEF with SDS-PAGE [53, 57, 59]. Combining techniques to fractionate samples is ideally the best manner to study the proteome, but by multiplying the fractions it is time-consuming. Therefore, it cannot realistically be envisaged in most cases in which a high number of technical and biological replicates will have to be analyzed.

The Luminex™ 100 system is a powerful tool for simultaneous quantification and analysis in multiple expression profile monitoring of disease-responsive proteins. It is a multiplex fluorescence flow microsphere system capable of simultaneously analyzing multiple analytes within a sample, including proteins and nucleic acids [61, 62]. For protein analysis, antibodies to specific proteins of interest are attached to microspheres (beads) that are internally dyed with red and orange fluorochromes emitting light of two different wavelengths. The ratio of the two fluorochromes can be varied and the system's digital signal processor can differentiate each bead set based on the fluorescent emission ratio. The Luminex™ system enables up to 100 different bead sets to be distinguished and, thus, by labeling each bead set with an antibody, up to 100 proteins can theoretically be analyzed simultaneously. Using a capture sandwich assay format, a reporter antibody labeled with a third fluorochrome binds to the analyte captured on the beads. Upon excitation by a laser, the intensity of the fluorescent signal emitted by the reporter fluorochrome is proportional to the amount of protein bound to the bead. The microspheres are analyzed by a flow cytometer that detects the intensities of all three fluorescent signals for each bead that passes. The multiplexed assay format does not lose accuracy and sensitivity in comparison with single analyte assays. The Luminex™ system is flexible, scalable, clinically approved, and allows the development of new specific tests. The technology is therefore promising for biomarker discovery and for clinical applications at the end of the proteomic biomarker discovery pipeline.

Protein arrays are also very interesting assays either to explore biological systems or to design specific chips for clinical applications. They are solid-phase ligand-binding systems using proteins immobilized on various surfaces such as glass, membranes, mass spectrometer plates, beads or particles. An antibody array is a very common format of protein array in which antibodies are covalently bound to surfaces and used to detect target molecules in complex samples using direct labeling or dual antibody sandwich assays [63, 64]. Antibody arrays can also be employed for the profiling of glycan structures on proteins [65]. In this case, antibody arrays capture multiple specific proteins directly from biological samples, and lectin- and glycan-binding antibodies probe the levels of specific glycans on the captured proteins.

The current limitation of Luminex™ and protein array technologies for protein analysis remains the difficulty of developing large-scale multiplexed arrays due to variable qualities of antibodies, cross-reactivities between different antibodies, and the huge dynamic range of protein concentrations in serum and plasma. Moreover, the expression and purification of immobilized proteins remains a time-consuming and laborious process.

One of the central applications of serum/plasma proteome analysis is to discover biomarkers for human cancer detection. Proteomics tools have been used extensively for the identification of serum biomarkers of different types of cancers such as prostate cancer, colorectal cancer, brain tumor, breast cancer, ovarian cancer, lung cancer, liver cancer, pancreatic cancer and gastric cancer, as reviewed by Hu et al. [24] and Ray et al. [66]. Antibody array for serum profiling has been well demonstrated in human cancers such as prostate, pancreatic and lung cancers [67–69]. A specific array for pancreas, liver, ovarian and gastrointestinal tumors has been applied to the detection of these different tumors in 15,867 individuals, revealing 16 cases afterwards confirmed as having cancers [70]. Cancer-associated glycan alterations of proteins in the serum of pancreatic cancer patients have been revealed by using antibody array for glycan structures [65].

Another interesting application of serum/plasma profiling using proteomics is the detection of autoimmune diseases, such as systemic lupus erythematosus, rheumatoid arthritis, multiple sclerosis, Crohn's disease, ulcerative colitis (bowel diseases). Serum/plasma profiling has also led to proposal of candidate biomarkers for infectious diseases, such as tuberculosis, severe acute respiratory syndrome, hepatitis and leprosy, and for other human diseases such as stroke, heart diseases, non-alcoholic fatty liver disease, liver fibrosis, liver cirrhosis, diabetic nephropathy, Down syndrome, sarcoidosis, failure of blood–brain or blood-cerebrospinal barriers, and schizophrenia. All these proteomic-based approaches for the identification of biomarkers have been well reviewed by Hu et al. [24] and more recently by Ray et al. [66].

Despite many challenges, serum and plasma proteomics, and more widely body fluid proteomics, is still one of the most promising approaches to disease biomarker study. However, biomarker discovery requires a robust statistical strategy for sample size calculation, data analysis and prediction modeling. Determining sample size is an important issue because too small size sampling may lead to inaccurate results. A major difficulty in regard to statistical prediction modeling is that there is no single pathognomonic feature that characterizes many human diseases such as cancers and systematic diseases, and particularly radiation diseases. Models built with several biomarkers will therefore often increase the statistical power of the study [71–73].

4.4 Current Knowledge About Proteins Differentially Expressed in Serum and Plasma Following Irradiation

4.4.1 Overview

Three serum/plasma biomarkers have been clinically validated and can be used in the diagnostic evaluation of radiation exposure (amylase), or in the evaluation of radiation injury to the bone marrow (Flt-3-ligand) or to the small bowel (citrulline). Amylase is an indicator of radiation damage to the parotid gland [74, 75], Flt3-ligand (Flt3-L) is an indicator of bone marrow damage [76] and the amino acid citrulline is used as a physiologic marker for epithelial radiation-induced small bowel damage [77]. These biomarkers are, however, not sufficiently sensitive for a fast and accurate triage, especially if individuals are assessed more than 48 h after exposure. Moreover, there is no approach today that indicates the diagnosis or the prognosis of radiation tissue toxicity for tissues other than bone marrow and intestine.

We recently reviewed current knowledge on protein biomarkers for radiation exposure and concluded that proteomics and multivariate approaches are essential as new investigation tools for the discovery of multiparametric biomarkers either for radiation exposure or for radiation tissue toxicity [78]. In the quest for radiation biomarker discovery, the overall purpose is to identify a set of variables that altogether allow classification of individual samples. It is essential to differentiate exposed from non-exposed individuals and between samples from people suffering from a radiation injury and samples from people who are not. Univariate variable assessment is simple to perform, but in most complex biological conditions it is unusual to find a single variable that is both sensitive and specific for a given exposure condition. For example, a number of acute phase reactant proteins are increased in serum/plasma or tissue of animals following local skin irradiation [79–84], but these are generally not considered to be specific to radiation-induced skin injury. Using multiple variables simultaneously is a common method to overcome the limitation of single markers. Multivariable analysis is therefore an important component of any study aimed at identifying biomarkers. Multivariate analysis can be divided into two general classes: supervised and unsupervised. Supervised learning requires that information regarding the outcome (class assignment) is available and is useful in developing a diagnostic tool during the development of a classification model. Unsupervised learning does not have such a requirement and is more powerful for data exploration.

We review here all published studies to our knowledge that employed proteomics methods to discover serum or plasma proteins related to radiation exposure or toxicity. Review of the literature shows that two main goals appear to be pursued when searching for radiation biomarkers. The first goal concerns radiological accidents or malevolent use of a radiological or nuclear device, with the aim of estimating the level of radiation exposure of an individual and if possible specifying exactly the dose of exposure (biodosimetry). In this case, the estimation of an overall dose is important for the rapid triage of exposed and non-exposed individuals. The second goal is related to the assessment and prediction of normal tissue toxicity before, during and after a treatment by radiotherapy. On the one hand, the small percentage of patients who are hypersensitive to radiation should be identified before the treatment. This may allow delivery of higher radiation doses to patients who are not hypersensitive, thus improving local tumor control and patient survival, for example in non-small cell lung cancer (NSCLC) [85]. On the other hand, it may allow personalized measures of clinical response to radiotherapy and rapid response measures for populations exposed to high doses of radiation. Further refinement of the methods may lead to personalized molecular medicine approaches directly applicable to future clinical trials and patient care, with the aim of escalating the radiation dose to tumors while avoiding toxicity for maximized therapeutic gain.

We identified 11 studies (10 articles and a patent application) in which proteomics was used to discover new radiation biomarkers for radiation exposure or tissue toxicity [84, 86–95]. In addition, 6 studies used the multiassay of proteins—using the Luminex™ system to assay several cytokines in most cases—to study serum/plasma proteins following irradiation or to discover new protein biomarkers of exposure or tissue toxicity [96–101]. Table 4.1 lists and summarizes studies that have looked for serum/plasma protein variations following radiation exposure, and emphasizes the principal proteins studied or identified in these studies. Of these 17 studies, 5, including the patent application, were interested in the discovery of biomarkers of radiation exposure [89, 91, 93, 95, 100], while the 12 others looked for biomarkers of normal tissue toxicity, generally as a consequence of a treatment by radiotherapy [84, 86–88, 90, 92, 94, 96–99, 101]. In the latter case, eight papers were on lung toxicity (i.e. pneumonitis or lung fibrosis) during and after radiation therapy [86, 87, 90, 94, 96, 98, 99, 101], one paper was on skin radiation injury [84], one on injury to several tissues (acute gastrointestinal and genitourinary toxicities) following radiotherapy for the treatment of prostate cancer [97], one on liver radiation injury [88], and one on acute mucosal reaction associated with radiotherapy [92].

4.4.2 Radiation Exposure

Biomarkers of radiation exposure have been investigated with different proteomics methods

Table 4.1 Serum/plasma protein profiling following irradiation

Context/application	Species	Exposure	Radiation dose	Sample	Sample collection timing	Method	Principal protein identified	Year	Reference
Radiation exposure	Radiation exposure, triage, biomarkers (RT exposure)	Human	Various sites	1.5–86.4 Gy	Serum	Before and during the latter part of the course of RT	SELDI-TOF-MS	Spectral components	2006	[89]
				6–62 Gy	Serum		1D-GE slices/LC-MS/MS	23 protein fragments/peptides including IL-6		
Radiation exposure	Radiation exposure, triage, biomarkers	Mouse	Whole body	3 Gy	Plasma	Days 2 and 7 post-IR	2D-GE/MS	Clusterin, gelsolin, kininogen, and alpha-2-HS-glycoproteins, inflammatory proteins	2009	[93]
Radiation exposure	Radiation exposure, triage, biomarkers (patent)	Mouse	Head	3 Gy	Tongue and buccal swab (serum/plasma application)	15 and 30 min post-IR	MALDI-TOF-MS/MS and MALDI imaging	LTGF-β and various proteins or peptides	2009	[91]
Radiation exposure	Radiation exposure, triage, biomarkers	Mouse Human	Whole body Cell lines	1, 2, 3 Gy 10 Gy	Cell lines	1 h post-IR 1, 2, and 3 days post-IR	High-throughput antibody-based assays ELISA	IL-6	2010	[100]
Radiation exposure	Radiation exposure, triage, biomarkers	Mouse	Whole body	5 Gy	Plasma	6 h post-IR		IL-6, IL1β, TNFα, TGFβ		
Radiation exposure	Radiation exposure, triage, biomarkers (RT exposure)	Human	Larynx	51–72 Gy	Serum	Before the start, 2 weeks after the start, and 4–6 weeks after the end of RT	MALDI-TOF MS (LMW)	Spectral components (41 peptides)	2011	[95]

(continued)

Table 4.1 (continued)

Context/application		Species	Exposure	Radiation dose	Sample	Sample collection timing	Method	Principal protein identified	Year	Reference
Radiation tissue toxicity	Lung toxicity, RT, NSCLC, prediction, dose escalation	Human	Lung	66–72 Gy	Plasma	Before the initiation of RT	Multiplex suspension bead array system	IL-8	2005	[98]
Radiation tissue toxicity	Lung toxicity, RT, NSCLC, pneumonitis, fibrosis, prediction	Human	Lung	Not specified	Plasma	Before the start, 2 weeks and 4 weeks after the start of RT	Multiplex suspension bead array system	IL-1 receptor antagonist (IL-1ra), ratio of IL-1ra to IL-1β, MCP1, IP10	2008	[99]
Radiation tissue toxicity		Mouse	Thorax	Not specified	Plasma	Not specified	SELDI-TOF MS	Spectral components		
Radiation tissue toxicity	Lung toxicity, RT, NSCLC, pneumonitis, prediction	Human	Lung	32 Gy, 60–66 Gy	Plasma	Before and weekly during RT, during follow-up (1, 3, 6, 9 months after RT), and at the onset of pneumonitis	ELISA	TNF-α, IL-1β, IL-6 and TGF-β1	2008	[101]
Radiation tissue toxicity	Lung toxicity, RT, NSCLC, pneumonitis, prediction	Human	Lung	70 Gy	Serum	Before RT, midtreatment, immediately after RT, and at a 3 and 6-month follow-up appointments	Data-mining methods based on machine learning	Spectral components	2009	[94]
Radiation tissue toxicity		Human	Lung	70 Gy	Serum	Before RT and last available follow-up	Shotgun proteomics (LC-MS/MS)			
Radiation tissue toxicity	Lung toxicity, prediction	Mouse	Whole lung	12 Gy	Serum, BALF	3, 6, 12, 24 h to 1 week post-IR	Multiplex suspension bead array system	G-CSF, IL-6, KC, MCP-1, IP-10	2009	[96]

Radiation tissue toxicity	Lung	Lung toxicity, RT, NSCLC, pneumonitis, prediction	Human	Lung	64 Gy	Plasma	Before RT, at 2, 4, 6 weeks during RT, and 1 and 3 months after RT	ExacTag labeling, RP-HPLC-nano-LC-ESI-MS/MS	C4b-binding protein alpha chain, complement C3, and vitronectin	2010	[86]
Radiation tissue toxicity	Lung	Lung toxicity, RT, NSCLC, pneumonitis, prediction	Human	Lung	67 Gy	Plasma	Before RT	ExacTag labeling, RP-HPLC-nano-LC-ESI-MS/MS	C4b-binding protein alpha chain and vitronectin	2011	[87]
Radiation tissue toxicity	Lung	Lung toxicity, RT, NSCLC, pneumonitis, prediction	Human	Lung	Not specified	Plasma	Before, during, at the end of RT and at 3 and 6-months follow-up	Graph-based scoring function and LC-MS/MS	Alpha-2-macroglobulin, alpha-1-antichymotrypsin, complement factor B, etc.	2011	[90]
Radiation tissue toxicity	Other organs	Skin toxicity, triage, RT	Mouse	Skin	40 Gy	Serum	1, 5, 14, 21 and 33 days post-IR	2D-DIGE/MALDI-TOF-MS	20 proteins (inflammation, acute phase response, coagulation)	2007	[84]
Radiation tissue toxicity	Other organs	Gastrointestinal and genitourinary toxicities, RT, prostate cancer, acute toxicity prediction	Human	Prostate	66 and 78 Gy	Serum	At 2 pretreatment appointments, at the 5th or 10th fraction during IMRT and at the last day of RT	Multiplex suspension bead array system	IFN-γ IL-6, IL-2 and IL-1	2009	[97]
Radiation tissue toxicity	Other organs	Liver toxicity, RT, hepatocellular carcinoma, hepatic toxicity	Rat	Liver	10 Gy	Serum	3 weeks post-IR	2D-GE/MS	About 60 proteins including heparanase precursor and hepatocyte growth factor	2010	[88]
Radiation tissue toxicity	Other organs	Acute mucosal toxicity, RT, head and neck cancer, acute mucosal reaction	Human	Head and neck	52–76 Gy	Plasma	Before the start of RT	MALDI-TOF MS (LMW)	Spectral components	2011	[92]

and models (Table 4.1). The potential use of proteomics has been explored in patients with cancer before and during radiotherapy in an effort to discover clinical biomarkers of radiation exposure [89]. SELDI-TOF-MS was used to generate high-throughput proteomic profiles of serum samples. The proteomic profiles were analyzed for unique biomarker signatures using supervised classification methods. In the same work, MS-based protein identification was done on patient serum to identify specific protein fragments that are altered following radiation exposure. Computer-based analyses of the SELDI-TOF protein spectra could distinguish unexposed from radiation-exposed patient samples with 91–100% sensitivity and 97–100% specificity using various classifier models. The method also showed an ability to distinguish high from low dose-volume levels of exposure with a sensitivity of 83–100% and specificity of 91–100%. Using direct identity techniques of albumin-bound peptides, 23 protein fragments and/or peptides were uniquely detected in the radiation exposure group, including an IL-6 precursor protein. This is the most interesting study that has been published since 2006 because it was performed in a rather large cohort of human patients, whatever the irradiated site and the dose received, using two complementary methods, and by investigation of the albumin-bound proteome. However, to date this promising study was not followed up by the publication of other papers. It would have been interesting, in particular, to measure uniquely detected protein fragments and/or peptides in the blood of the patients, and to identify spectral components of the SELDI-TOF analysis.

A recent paper has reported the use of a similar strategy to look for radiation exposure biomarkers in the serum of patients treated by radiotherapy for larynx cancer [95]. The LWM fraction of the serum proteome (2,000–13,000 Da) was analyzed by MALDI-TOF-MS. Mass profiles of serum samples collected several weeks after the end of the treatment revealed changes compared with earlier time-points. The authors concluded that features of the serum proteome are potentially applicable as a retrospective marker of exposure to ionizing radiation and could be applicable in radiation dosimetry. However, the potential association of dose and serum feature of this preliminary study lacked strong statistical significance, likely because of the small number of patients analyzed. Furthermore, this work only identified spectral mass components, and not proteins, which means that this study is not easily applicable.

Protein biomarkers of radiation exposure have also been sought using proteomics tools in animal models. Rithidech et al. exposed mice to a whole-body dose of 3 Gy in an attempt to find differentially expressed plasma proteins at days 2 and 7 post-irradiation using 2D-GE coupled to mass spectrometry [93]. The majority of proteins with significantly altered expression levels were acute phase proteins (APPs), suggesting an inflammatory response. This inflammatory response could continue from day 2 to day 7 post-exposure. Changes in expression of certain proteins (gelsolin, clusterin, HMW kininogen, α-2-HS glycoproteins) were detected at both time-points and could be dependent on radiation dose and time post-exposure. However, validation is needed to postulate that these changes in expression levels may be indicative of radiation exposure. In addition, a patent was filed in 2009 in the USA covering the use of novel radiation-associated markers to assess radiation exposure [91]. The invention relates to new protein biomarkers that may be present after exposure to ionizing radiation and to methods of assessing exposure to ionizing radiation as well as diagnostic tests and kits for evaluating exposure to ionizing radiation. The radiation-associated markers may be one or more of albumin, LTGF-β, or any protein or peptide listed in any one of the Tables provided in the patent application (read more on www.faqs.org/patents/app/20090289182#ixzz1l7qHYDDu). The invention also provides methods of assessing exposure to ionizing radiation by determining the presence of one or more radiation-associated markers. Differentially expressed proteins have been searched for in tongue and buccal swabs of mice after 1–3 Gy irradiation of the head or the whole body, 15 min, 30 min and 1 h post-irradiation. MALDI-TOF-

MS/MS and MALDI imaging generates lists of peptides of which some identified proteins like albumin and TGF-β. The inventor claims that the levels of these proteins and peptides may be evaluated in saliva, a buccal swab, amniotic fluid, plasma, serum, urine or blood. The patent covers a broad panel of applications like the amount of radiation therapy that has been delivered to a particular tissue, kits for assessing the likelihood of significant damage, death, illness or medical complications after exposure to elevated levels of ionizing radiation, etc.

High-throughput antibody-based assays have been performed in an effort to look for differentially expressed proteins after irradiation. Partridge et al. [100] systematically examined the most promising candidate antigens of the 260 proteins that had been reported to be responsive to radiation [102]. They used 10 cultured human cells (normal cell lines in preference to tumor cell lines), including hematopoietic cells, to detect changes in expression of 12 secreted proteins using ELISA as the detection system. All cell lines were exposed to moderate to high levels of γ-radiation (2, 5 or 10 Gy) and the culture media were tested for secreted proteins. Subsequently, they assessed whether the cells increased the expression of these proteins after irradiation, or if cells that were initially negative could be induced to express the protein above the threshold of detection, identifying only IL-6 as a significantly secreted cytokine. After this systematic in vitro screen, they measured changes at 6 h after irradiation in the level of a subset of these candidate proteins in plasma from C57BL/6 mice whole-body irradiated with a single radiation dose of 5 Gy. They found that IL-6, IL-1β, TNF-α, and TGF-β levels were altered after radiation exposure. This work highlighted the potential for IL-6 as a marker for an immunoassay-based high-throughput biodosimeter. The next step of this work would be to assess the ability of IL-6 to serve as a radiation biomarker at different timepoints after exposure to different doses of irradiation. It will then be critical to confirm efficacy of these assays in larger populations and in human patient samples.

4.4.3 Radiation Tissue Toxicity

The lung is the most investigated organ for the study of radiation tissue toxicity because radiotherapy is extensively used for the treatment of lung cancer. Most studies have used human samples, serum or plasma, collected before, in the course of, or after radiotherapy, and have looked for predictive biomarkers of lung toxicity or markers of radiation-induced lung toxicity (Table 4.1).

In some studies, the multiplex suspension bead array system (Luminex™ system) was used to simultaneously profile multiple proteins induced by ionizing radiation. In 2005, Hart et al. evaluated the level of 17 cytokines in the plasma of patients before the initiation of radiotherapy [98]. Patients with lower levels of plasma IL-8 before radiation therapy might be at increased risk for developing lung injury. Significant correlations were not found for any other cytokine in this study, including TGF-β, which appeared to correlate with lung toxicity when evaluated during radiotherapy [103]. However, further studies are necessary to determine whether IL-8 levels are predictive of radiation-induced lung injury in a prospective trial and whether this marker might be used to determine patient eligibility for dose escalation. Designed to predict radiation lung toxicity, blood biomarkers have been used as a potential strategy to individualize thoracic radiation therapy [99]. The authors reviewed data associated with cytokines and updates of proteomic and genetic polymorphisms in radiation lung toxicity. They demonstrated significant values of cytokines such as TGF-β1, IL-6, KL-6, surfactant proteins, and IL-1ra in predicting radiation-induced lung toxicity. According to the authors, biomarkers or models that can accurately predict radiation-induced lung damage at an early stage, before completion of chemoradiotherapy, could allow physicians to monitor and customize the treatment to each patient. In their review, Kong et al. also presented their own unpublished results on the assessment of 29 cytokines in the plasma of patients for correlation with radiation-induced lung toxicity. The cytokines that generated a radiation damage protective effect such as IL1ra and the ratio of IL-1ra to IL-1β were significantly higher during the course of the treatment for patients without lung toxicity. The cytokines stimulating radiation damage such as MCP-1 and IP-10 often have more significant radiation-induced elevation at 4 weeks during the course of radiation in patients with lung toxicity several months after completion of treatment. Another study profiled 22 cytokines in the lung tissue homogenates, BALF, and serum from 3, 6, 12, 24 h to 1 week after 12 Gy whole-lung irradiation of mice sensitive (C57BL/6) and resistant (C3H) to lung fibrosis [96]. Radiation induced earlier and greater temporal changes in multiple cytokines in the pulmonary fibrosis-sensitive mice, particularly for G-CSF, IL-6, KC, MCP-1, and IP-10. In addition, a positive correlation between serum and tissue levels suggested that blood may be used as a surrogate marker for tissue. An additional study investigated the prognostic value of TNF-α, IL-1β, IL-6 and TGF-β1 plasma levels in predicting radiation pneumonitis [101]. The results of this study did not confirm that cytokine plasma levels, either absolute or relative values, identify patients at risk for radiation pneumonitis. In contrast, the clear correlations of IL-6 and TGF-β1 plasma levels with cytokine production in corresponding tumor biopsies and with the individual tumor responses suggest that the tumor is the major source of circulating cytokines in patients receiving radiation therapy for advanced NSCLC. This study indicates that further studies are necessary to identify reliable predictors of adverse radiotherapy effects.

Several studies have used the potential of gel-free proteomics to investigate biomarkers of early or late injury to the lung in the course of or following radiotherapy in humans. I. El Naqa and collaborators have coupled bioinformatics methods and shotgun proteomics (LC-MS/MS) to identify biomarkers of radiation-induced lung inflammation. In 2009, their preliminary results identified mass spectral components related to radiation pneumonitis [94]. Their proteomics strategy could have identified biomarkers relevant to the inflammation response. According to their paper, they are currently investigating incorporation of these biomarkers into their existing dose-volume model of radiation pneumonitis to improve its predictive power and potentially to demonstrate its feasibility for individualization of radiotherapy of NSCLC patients. In 2011, they proposed a bioinformatics approach for biomarker identification in radiation-induced lung inflammation from limited proteomics data [90]. On the basis of the proposed methodology, α-2-macroglobulin (α2M) was ranked as the top candidate protein. They also found many other proteins, most of them related to inflammation, such as complement components, hemopexin and α-1-antichymotrypsin. The authors suggested that the methodology based on longitudinal proteomics analysis and a novel bioinformatics ranking algorithm is a potentially promising approach for the challenging problem of identifying relevant biomarkers in sample-limited clinical applications. Cai et al. have also performed shotgun proteomics to identify biomarkers of radiation pneumonitis during the course of radiation therapy, or to predict, before the start of the treatment, patients at risk of developing lung toxicity. They used a multiplexed quantitative proteomics approach involving ExacTag labeling, RP-HPLC and nano-LC-MS/MS. They showed that C4b-binding protein alpha chain, complement C3, and vitronectin had significantly higher expression levels in patients with lung injury compared with patients without lung injury, based on both the data sets of radiotherapy start to 3 months post-radiotherapy and radiotherapy start to the end of radiotherapy [86]. In another paper, they also identified C4b-binding protein alpha chain and vitronectin, with the same methodology, as markers predictive of patients at risk and found that the inflammatory response probably plays an important role [87].

We developed in our laboratory a proteomic approach using an animal model in mice to look for biomarkers of skin damage after local irradiation [84]. In this model, mice developed reproducible clinical signs on the skin ranging from erythema to necrosis. Global proteomics approaches were used to look for serum proteins altered in expression during the early post-irradiation phase. 2D-DIGE coupled with mass spectrometry was used to investigate proteins altered in expression and/or post-translational modifications in serum. Proteome changes were monitored from 1 day to 1 month post-irradiation, at a dose of 40 Gy. About 60 proteins (including some isoforms and likely post-translational variants), representing 20 different proteins that exhibited significant and reproducible kinetic expression changes, were identified using mass spectrometry and database searches. Several proteins, down- or up-regulated from day 1, could prove to be good candidates to predict the progression of a skin lesion such as necrosis. Identified proteins are mainly related to inflammation and the acute phase response (e.g. α1-antitrypsin, ApoA-1, complement components, haptoglobin, hemopexin), coagulation (e.g. antithrombin III, thrombospondin, kininogen I, murinoglobulin 1) and are known to play a role in wound repair. Changes in APPs of irradiated mice, as pointed out by Chen et al. [82], may serve as antioxidant and anti-inflammatory factors to reduce the damage to skin. Variation in the levels of proteins involved in the coagulation system may reflect the severe vascular injury caused by irradiation to the vasculature [16], and particularly to the endothelium. We also observed shifts in isoelectric point (p*I*) of several spot trains, revealing potential post-translational changes which could also serve as indicators of irradiation or as predictors of lesion severity. In our current work, we have shown that high doses of irradiation of the skin led to changes in the number of glycan structures carried by serum proteins [114]. Using a large-scale quantitative mass spectrometry-based glycomic approach, we performed a global analysis of glycan structures of serum proteins from non-irradiated and locally irradiated mice exposed to different high doses of γ-rays (20, 40 and 80 Gy). Unsupervised descriptive statistical analyses (principal component analysis) using quantitative glycan structure data allowed us to discriminate between uninjured/slightly injured animals and animals that developed severe lesions. Decisional statistics showed that several glycan families were down-regulated while others increased, and that particular structures were statistically significantly changed in the serum of locally irradiated mice. Our results suggest for the first time a role of serum protein glycosylation in the response to irradiation. These protein-associated glycan structure changes may signal radiation exposure or effects. Overall, our results indicate that the serum proteome and glycome contents are readily modified after local irradiation of the skin, showing that their investigation can be of great relevance in identifying diagnostic or prognostic bioindicators. Our goal is now to define sets of serum proteins whose variations could help to predict the severity of the upcoming lesion.

Other tissue toxicities have also been investigated in the context of normal tissue damage initiated by radiotherapy (Table 4.1). Multiplex suspension bead arrays (Luminex™ system) were used to investigate the serum of patients treated for prostate cancer with IMRT and to search for positive correlations with gastrointestinal and genitourinary toxicity [97]. The results showed a significant increase in IFN-γ and IL-6 during IMRT. Increased IL-2 and IL-1 expression were associated with increased probability of acute gastrointestinal and genitourinary toxicity, respectively. Also, predictive biomarkers of acute mucosal toxicity were sought using mass spectrometry in the LMW plasma proteome of patients treated for head and neck cancer [92]. Spectral components recorded in plasma samples were used to build classifiers that discriminated patients from healthy individuals with about 90% specificity and sensitivity. Four spectral components were identified whose abundances correlated with a maximal intensity of the acute reaction. Several spectral components whose abundances correlated with the rate of DNA repair in irradiated lymphocytes were also detected. Additionally, a more rapid escalation of an acute reaction was correlated with a higher level of unrepaired damage assessed by the comet assay. It was concluded that the plasma proteome could be considered as a potential source of predictive markers of acute reaction in patients with head and neck cancer treated with radiotherapy. Animal models have also been used to find biomarkers of tissue toxicity in the context of radiotherapy. Radiation hepatic toxicity was modeled in cirrhotic rats to mimic the treatment of hepatocellular carcinoma [88]. Liver tissue and serum were analyzed using 2D-GE and Q-TOF-MS. Identified proteins were validated using western blotting. Histological examination showed that the degree of hepatic fibrosis was increased by radiation in liver cirrhosis. Fibrosis was associated with decreased proliferation of cell nuclear antigen and increased apoptosis. Proteomic analysis of liver tissue and serum identified 60 proteins which showed significant differences in expression between non-exposed and exposed groups. Among these, an increase of heparanase precursor and decrease of hepatocyte growth factor were shown commonly in liver tissue and serum following radiation. Hepatic fibrosis increased following radiation in cirrhotic rats. These proteins might be useful in detecting and monitoring radiation-induced hepatic injury. Lastly, SELDI-TOF may be able to identify spectral components correlated with lung toxicity in strain mice exhibiting different sensitivities to lung injury [99]. This study appears to show significant differences, either up-regulated or down-regulated, in protein profiles in plasma between radiation-resistant C3H mice and radiation-sensitive C57BL/6 mice well before a histopathologically detectable difference occurred. However, these results should be considered with caution since they have yet to be corroborated.

4.4.4 Conclusions

Changes in the levels of various proteins are mostly consistent with the acute phase response to tissue injury and inflammation. For instance, α1-anti-trypsin, ApoA-I, complement components, contrapsin (a human α1-anti-chymotrypsin homologue in mouse), haptoglobin, hemopexin, murinoglobulin 1 and serum amyloid P-component have been identified in several studies following irradiation [84, 86, 87, 90, 93]. APPs increase (positive APPs) or decrease (negative APPs) following different injuries, reflecting the presence and intensity of inflammation [104]. Exposure to ionizing radiation has previously been shown to induce the production of APPs in rat [80, 81] and mouse [79, 105] tissues, in the extracellular protein content of murine bone marrow [82] and during radiotherapy [83]. The changes in APPs of irradiated mice may serve as antioxidant and anti-inflammatory factors to reduce the damage to skin [82], although the exact mechanism underlying the radiation-induced acute phase response remains unclear. Blood composition is modified after exposure to ionizing radiation through the immediate release by irradiated cells of cytokines and growth factors which stimulate neighboring cells or distant cells which in turn release proteins into the extracellular environment. IL-1, IL-6, IL-8, TGF-β, TNF-α and eotaxin are the major known cytokines involved in the response to ionizing radiation [19, 106]. Cytokines play a role in mediating the inflammation process, and stimulate or repress APP synthesis in the liver [104], assisting the repair of the tissue. Following localized radiation injury, this inflammation process is a complex and in part specific mechanism that is only beginning to be elucidated. In this regard, it is not surprising that differential expression of several cytokines in the serum and the plasma after irradiation has been identified using high-throughput antibody-based assay [96–101].

Apart from APPs, several proteins involved in the coagulation system also change in the serum after irradiation (α1-anti-trypsin, contrapsin, murinoglobulin 1, kininogen I, antithrombin III, thrombospondin 1 and Pzp). These variations may reflect the intense vascular injury caused by ionizing radiation to the vasculature [16], and particularly to the endothelium. A decrease of antithrombin III and murinoglobulin 1 (which belongs to alpha-2-macroglobulin), and an increase of kininogen I and thrombospondin 1 may reveal an activation of the coagulation system in response to endothelial damage.

Several other proteins also change in the serum and may account for different protective functions or pathophysiological processes. For example, we have identified [84] carboxylesterase, which is involved in the metabolism of xenobiotics and of natural substrates and may eliminate toxic metabolites released by damaged tissues [107], gelsolin, involved in the clearance of actin [108], and meprin A, both of which afford protection against the release of components during tissue injury, MHC I H2-Q10 α chain, which is involved in the immune system and in inflammation, and peroxiredoxin 2, a peroxidase with antioxidant activities which decreases in the extracellular protein content of murine bone marrow after total-body irradiation [82].

In conclusion, the many studies we have reviewed here clearly indicate that serum/plasma protein content is dynamically and reproducibly modified after exposure to ionizing radiation in various animal models and in clinical studies in human subjects. Physiologic response of the body to injury by ionizing radiation mostly involves acute phase response, coagulation and other cellular defense processes and helps our understanding of how an organism manages this highly complex response. Although it is questionable whether any of the discovered proteins are specific for radiation exposure and not just for injuries, the variations of expression of a set of these proteins, coupled with post-translational modifications, may appear to be highly specific to radiation exposure. However, more work is needed to validate these biomarkers, particularly for the prediction of radiation response. In particular, a correlation still needs to be made between biomarkers and the severity of the radiation reaction.

4.5 Concluding Remarks and Future Prospects: Molecular Networks as Sensors of Radiation Diseases?

Tissue responses to radiation injury are complex: damaging or protective, they probably result from collective and synergistic actions of many factors. The proteomic approach makes an important contribution to revealing an image of the pathological states of every tissue exposed to ionizing radiation in the body. The search for a unique biomarker of radiation exposure or tissue toxicity has the advantage of simplicity, but makes little sense from a biological viewpoint given that radiation-induced diseases are very heterogeneous. Irradiation injuries are caused by intrinsically disturbed cells, allowing huge molecular, cellular, tissular, immunologic and systemic changes, sometimes far from the site of irradiation, such as in the liver or in the bone marrow. Transcriptomics researchers are profiling thousands of gene transcripts to generate full patterns of information. We should be able to do the same with protein biomarkers in the field of radiation research. The relative abundance of thousands of different proteins, with their cleaved or modified form, is an image of ongoing physiological and pathological events. Also, it has been demonstrated that coherent networks of genes respond to genetic and environmental perturbations and in turn influence disease-associated traits [109], showing that diseases are probably emergent properties of networks rather than the result of single gene or protein expression changes. In the field of radiation research, considering that ionizing radiation causes disease [110, 111], we should thus be able to link gene and protein expression studies to genome-wide association studies (GWAS), which are now well proven to uncover genetic loci that affect disease risk or progression [112], and also to other omics studies like metabolomics. The idea of molecular networks as sensors has emerged recently [113] and should be applied to the field of radiation research. There is a need to develop a more comprehensive understanding of the whole-system physiology that responds to radiation injury so as to link molecular biology to clinical medicine, i.e. molecular states to physiological states. This is now possible in the light of reverse engineering of molecular networks that sense DNA and environmental perturbations and drive variations in physiological states associated with disease [113]. Cells include many thousands of DNAs, RNAs, proteins, which can be modified

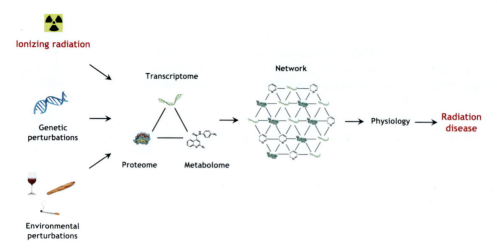

Fig. 4.3 Future prospects in radiation research: connecting molecular biology to radiation pathophysiology and radiation disease through molecular networks. The complexity of molecular biology, evaluated by the monitoring of DNA, RNA, proteins and metabolite variations, can drive a system view of radiation disease in which networks of interacting molecular entities are built to define pathophysiological states of the system. In this manner, a link can be established between molecular biology and clinical medicine (This figure has been adapted from Figs. 1 and 2 of the review by Eric E. Schadt [113])

at multiple sites, and metabolites interacting in complex pathways. Complex biological systems comprise in turn many types of cells which constitute very different organs that all interact in complex ways, giving rise to fully living systems. One goal is currently to model the extent of such interactions between molecular entities, between cells, and between organ systems. In the context of biological systems, a network can be viewed as a graphic model that represents the relationship between DNAs, RNAs, proteins, metabolites and phenotypes as disease states. The potentially millions of such interactions represented in a network define the overall connectivity structure of the network (Fig. 4.3). Variations of this network can in turn be considered as a sensor of a disease state. Since ionizing radiation causes disease states, molecular networks will be essential to study, detect or predict radiation effects.

Acknowledgements We are grateful to Electricité de France (Groupe Gestion Projet – Radioprotection) and the European Union (Seventh Framework Programme (FP7/2007–2013) under grant agreement n° 241536) which financially supports our work on biomarker discovery for molecular prognosis of tissue radiation toxicity.

References

1. Stone HB, McBride WH, Coleman CN (2002) Modifying normal tissue damage postirradiation. Report of a workshop sponsored by the Radiation Research Program, National Cancer Institute, Bethesda, 6–8 Sept 2000. Radiat Res 157:204–223
2. Stone HB, Coleman CN, Anscher MS et al (2003) Effects of radiation on normal tissue: consequences and mechanisms. Lancet Oncol 4:529–536
3. Rubin P, Johnston CJ, Williams JP et al (1995) A perpetual cascade of cytokines postirradiation leads to pulmonary fibrosis. Int J Radiat Oncol Biol Phys 33:99–109
4. Hong JH, Chiang CS, Tsao CY et al (1999) Rapid induction of cytokine gene expression in the lung after single and fractionated doses of radiation. Int J Radiat Biol 75:1421–1427
5. Haston CK, Zhou X, Gumbiner-Russo L et al (2002) Universal and radiation-specific loci influence murine susceptibility to radiation-induced pulmonary fibrosis. Cancer Res 62:3782–3788
6. Dorr W, Hendry JH (2001) Consequential late effects in normal tissues. Radiother Oncol 61:223–231
7. Gorin NC, Fliedner TM, Gourmelon P et al (2006) Consensus conference on European preparedness for haematological and other medical management of mass radiation accidents. Ann Hematol 85:671–679

8. Friesecke I, Beyrer K, Fliedner TM (2001) How to cope with radiation accidents: the medical management. Br J Radiol 74:121–122
9. Fliedner TM, Friesecke I, Beyrer K (2001) Medical management of radiation accident: manual on the acute radiation syndrome. METREPOL (European Commission Concerted Action). The British Institute of Radiology, Oxford
10. Shrieve DC, Loeffler JS (2011) Human radiation injury. Lippincott Williams & Wilkins, Philadelphia
11. Steel GG (2001) The case against apoptosis. Acta Oncol 40:968–975
12. Valerie K, Yacoub A, Hagan MP et al (2007) Radiation-induced cell signaling: inside-out and outside-in. Mol Cancer Ther 6:789–801
13. Fu XL, Huang H, Bentel G et al (2001) Predicting the risk of symptomatic radiation-induced lung injury using both the physical and biologic parameters V(30) and transforming growth factor beta. Int J Radiat Oncol Biol Phys 50:899–908
14. Chen Y, Rubin P, Williams J et al (2001) Circulating IL-6 as a predictor of radiation pneumonitis. Int J Radiat Oncol Biol Phys 49:641–648
15. Chen Y, Williams J, Ding I et al (2002) Radiation pneumonitis and early circulatory cytokine markers. Semin Radiat Oncol 12:26–33
16. Fajardo LF, Berthrong M (1988) Vascular lesions following radiation. Pathol Annu 23:297–330
17. Paris F, Fuks Z, Kang A et al (2001) Endothelial apoptosis as the primary lesion initiating intestinal radiation damage in mice. Science 293:293–297
18. Hauer-Jensen M, Kong FM, Fink LM et al (1999) Circulating thrombomodulin during radiation therapy of lung cancer. Radiat Oncol Investig 7:238–242
19. Barcellos-Hoff MH (1998) How do tissues respond to damage at the cellular level? The role of cytokines in irradiated tissues. Radiat Res 150:S109–S120
20. Brush J, Lipnick SL, Phillips T et al (2007) Molecular mechanisms of late normal tissue injury. Semin Radiat Oncol 17:121–130
21. Fajardo LF (2005) The pathology of ionizing radiation as defined by morphologic patterns. Acta Oncol 44:13–22
22. Rodemann HP, Bamberg M (1995) Cellular basis of radiation-induced fibrosis. Radiother Oncol 35:83–90
23. Corthals GL, Wasinger VC, Hochstrasser DF et al (2000) The dynamic range of protein expression: a challenge for proteomic research. Electrophoresis 21:1104–1115
24. Hu S, Loo JA, Wong DT (2006) Human body fluid proteome analysis. Proteomics 6:6326–6353
25. Sharma M, Halligan BD, Wakim BT et al (2008) The urine proteome as a biomarker of radiation injury. Proteomics Clin Appl 2:1065–1086
26. Anderson NL, Anderson NG (2002) The human plasma proteome: history, character, and diagnostic prospects. Mol Cell Proteomics 1:845–867
27. Thadikkaran L, Siegenthaler MA, Crettaz D et al (2005) Recent advances in blood-related proteomics. Proteomics 5:3019–3034
28. Lundblad RL (2005) Considerations for the use of blood plasma and serum for proteomic analysis. Internet J Genomics Proteomics 1:1–12
29. Misek DE, Kuick R, Wang H et al (2005) A wide range of protein isoforms in serum and plasma uncovered by a quantitative intact protein analysis system. Proteomics 5:3343–3352
30. Omenn GS, States DJ, Adamski M et al (2005) Overview of the HUPO Plasma Proteome Project: results from the pilot phase with 35 collaborating laboratories and multiple analytical groups, generating a core dataset of 3020 proteins and a publicly-available database. Proteomics 5:3226–3245
31. Tammen H, Schulte I, Hess R et al (2005) Peptidomic analysis of human blood specimens: comparison between plasma specimens and serum by differential peptide display. Proteomics 5:3414–3422
32. Bjorhall K, Miliotis T, Davidsson P (2005) Comparison of different depletion strategies for improved resolution in proteomic analysis of human serum samples. Proteomics 5:307–317
33. Fountoulakis M, Juranville JF, Jiang L et al (2004) Depletion of the high-abundance plasma proteins. Amino Acids 27:249–259
34. Fujii K, Nakano T, Kawamura T et al (2004) Multidimensional protein profiling technology and its application to human plasma proteome. J Proteome Res 3:712–718
35. Wang YY, Cheng P, Chan DW (2003) A simple affinity spin tube filter method for removing high-abundant common proteins or enriching low-abundant biomarkers for serum proteomic analysis. Proteomics 3:243–248
36. Anderson NL, Polanski M, Pieper R et al (2004) The human plasma proteome: a nonredundant list developed by combination of four separate sources. Mol Cell Proteomics 3:311–326
37. Rose K, Bougueleret L, Baussant T et al (2004) Industrial-scale proteomics: from liters of plasma to chemically synthesized proteins. Proteomics 4:2125–2150
38. Farrah T, Deutsch EW, Omenn GS et al (2011) A high-confidence human plasma proteome reference set with estimated concentrations in PeptideAtlas. Mol Cell Proteomics 10(M110):006353
39. Deutsch EW, Eng JK, Zhang H et al (2005) Human plasma PeptideAtlas. Proteomics 5:3497–3500
40. Tirumalai RS, Chan KC, Prieto DA et al (2003) Characterization of the low molecular weight human serum proteome. Mol Cell Proteomics 2:1096–1103
41. Hu L, Ye M, Zou H (2009) Recent advances in mass spectrometry-based peptidome analysis. Expert Rev Proteomics 6:433–447

42. Geho DH, Liotta LA, Petricoin EF et al (2006) The amplified peptidome: the new treasure chest of candidate biomarkers. Curr Opin Chem Biol 10:50–55
43. Mehta AI, Ross S, Lowenthal MS et al (2003) Biomarker amplification by serum carrier protein binding. Dis Markers 19:1–10
44. Tian Y, Zhang H (2010) Glycoproteomics and clinical applications. Proteomics Clin Appl 4:124–132
45. Arnold JN, Saldova R, Galligan MC et al (2011) Novel glycan biomarkers for the detection of lung cancer. J Proteome Res 10:1755–1764
46. Arnold JN, Saldova R, Hamid UM et al (2008) Evaluation of the serum N-linked glycome for the diagnosis of cancer and chronic inflammation. Proteomics 8:3284–3293
47. Bones J, Byrne JC, O'Donoghue N et al (2011) Glycomic and glycoproteomic analysis of serum from patients with stomach cancer reveals potential markers arising from host defense response mechanisms. J Proteome Res 10:1246–1265
48. Kyselova Z, Mechref Y, Al Bataineh MM et al (2007) Alterations in the serum glycome due to metastatic prostate cancer. J Proteome Res 6:1822–1832
49. Morelle W, Michalski JC (2007) Analysis of protein glycosylation by mass spectrometry. Nat Protoc 2:1585–1602
50. Pabst M, Altmann F (2011) Glycan analysis by modern instrumental methods. Proteomics 11:631–643
51. Petricoin EF, Ardekani AM, Hitt BA et al (2002) Use of proteomic patterns in serum to identify ovarian cancer. Lancet 359:572–577
52. Petricoin EF, Zoon KC, Kohn EC et al (2002) Clinical proteomics: translating benchside promise into bedside reality. Nat Rev Drug Discov 1:683–695
53. Pieper R, Gatlin CL, Makusky AJ et al (2003) The human serum proteome: display of nearly 3700 chromatographically separated protein spots on two-dimensional electrophoresis gels and identification of 325 distinct proteins. Proteomics 3:1345–1364
54. Cho SY, Lee EY, Lee JS et al (2005) Efficient prefractionation of low-abundance proteins in human plasma and construction of a two-dimensional map. Proteomics 5:3386–3396
55. Martosella J, Zolotarjova N, Liu H et al (2005) Reversed-phase high-performance liquid chromatographic prefractionation of immunodepleted human serum proteins to enhance mass spectrometry identification of lower-abundant proteins. J Proteome Res 4:1522–1537
56. Qin S, Ferdinand AS, Richie JP et al (2005) Chromatofocusing fractionation and two-dimensional difference gel electrophoresis for low abundance serum proteins. Proteomics 5:3183–3192
57. Tang HY, Ali-Khan N, Echan LA et al (2005) A novel four-dimensional strategy combining protein and peptide separation methods enables detection of low-abundance proteins in human plasma and serum proteomes. Proteomics 5:3329–3342
58. Wasinger VC, Locke VL, Raftery MJ et al (2005) Two-dimensional liquid chromatography/tandem mass spectrometry analysis of Gradiflow fractionated native human plasma. Proteomics 5:3397–3401
59. Horn A, Kreusch S, Bublitz R et al (2006) Multidimensional proteomics of human serum using parallel chromatography of native constituents and microplate technology. Proteomics 6:559–570
60. Sheng S, Chen D, Van Eyk JE (2006) Multidimensional liquid chromatography separation of intact proteins by chromatographic focusing and reversed phase of the human serum proteome: optimization and protein database. Mol Cell Proteomics 5:26–34
61. Nolan JP, Mandy FF (2001) Suspension array technology: new tools for gene and protein analysis. Cell Mol Biol (Noisy-le-Grand) 47:1241–1256
62. Nolan JP, Mandy F (2006) Multiplexed and microparticle-based analyses: quantitative tools for the large-scale analysis of biological systems. Cytometry A 69:318–325
63. Haab BB (2005) Antibody arrays in cancer research. Mol Cell Proteomics 4:377–383
64. Haab BB (2003) Methods and applications of antibody microarrays in cancer research. Proteomics 3:2116–2122
65. Chen S, Haab BB (2009) Analysis of glycans on serum proteins using antibody microarrays. Methods Mol Biol 520:39–58
66. Ray S, Reddy PJ, Jain R et al (2011) Proteomic technologies for the identification of disease biomarkers in serum: advances and challenges ahead. Proteomics 11:2139–2161
67. Gao WM, Kuick R, Orchekowski RP et al (2005) Distinctive serum protein profiles involving abundant proteins in lung cancer patients based upon antibody microarray analysis. BMC Cancer 5:110
68. Miller JC, Zhou H, Kwekel J et al (2003) Antibody microarray profiling of human prostate cancer sera: antibody screening and identification of potential biomarkers. Proteomics 3:56–63
69. Orchekowski R, Hamelinck D, Li L et al (2005) Antibody microarray profiling reveals individual and combined serum proteins associated with pancreatic cancer. Cancer Res 65:11193–11202
70. Sun Z, Fu X, Zhang L et al (2004) A protein chip system for parallel analysis of multi-tumor markers and its application in cancer detection. Anticancer Res 24:1159–1165
71. Anderson L (2005) Candidate-based proteomics in the search for biomarkers of cardiovascular disease. J Physiol 563:23–60
72. LaBaer J (2005) So, you want to look for biomarkers (introduction to the special biomarkers issue). J Proteome Res 4:1053–1059
73. Landers KA, Burger MJ, Tebay MA et al (2005) Use of multiple biomarkers for a molecular diagnosis of prostate cancer. Int J Cancer 114:950–956

74. Barrett A, Jacobs A, Kohn J et al (1982) Changes in serum amylase and its isoenzymes after whole body irradiation. Br Med J (Clin Res Ed) 285:170–171
75. Becciolini A, Giannardi G, Cionini L et al (1984) Plasma amylase activity as a biochemical indicator of radiation injury to salivary glands. Acta Radiol Oncol 23:9–14
76. Bertho JM, Demarquay C, Frick J et al (2001) Level of Flt3-ligand in plasma: a possible new bioindicator for radiation-induced aplasia. Int J Radiat Biol 77:703–712
77. Lutgens LC, Deutz NE, Gueulette J et al (2003) Citrulline: a physiologic marker enabling quantitation and monitoring of epithelial radiation-induced small bowel damage. Int J Radiat Oncol Biol Phys 57:1067–1074
78. Guipaud O, Benderitter M (2009) Protein biomarkers for radiation exposure: towards a proteomic approach as a new investigation tool. Ann Ist Super Sanita 45:278–286
79. Hong JH, Chiang CS, Campbell IL et al (1995) Induction of acute phase gene expression by brain irradiation. Int J Radiat Oncol Biol Phys 33:619–626
80. Magic Z, Matic-Ivanovic S, Savic J et al (1995) Ionizing radiation-induced expression of the genes associated with the acute response to injury in the rat. Radiat Res 143:187–193
81. Trutic N, Magic Z, Urosevic N et al (2002) Acute-phase protein gene expression in rat liver following whole body X-irradiation or partial hepatectomy. Comp Biochem Physiol C Toxicol Pharmacol 133:461–470
82. Chen C, Lorimore SA, Evans CA et al (2005) A proteomic analysis of murine bone marrow and its response to ionizing radiation. Proteomics 5:4254–4263
83. Cengiz M, Akbulut S, Atahan IL et al (2001) Acute phase response during radiotherapy. Int J Radiat Oncol Biol Phys 49:1093–1096
84. Guipaud O, Holler V, Buard V et al (2007) Time-course analysis of mouse serum proteome changes following exposure of the skin to ionizing radiation. Proteomics 7:3992–4002
85. Kong FM, Ten Haken RK, Schipper MJ et al (2005) High-dose radiation improved local tumor control and overall survival in patients with inoperable/unresectable non-small-cell lung cancer: long-term results of a radiation dose escalation study. Int J Radiat Oncol Biol Phys 63:324–333
86. Cai XW, Shedden K, Ao X et al (2010) Plasma proteomic analysis may identify new markers for radiation-induced lung toxicity in patients with non-small-cell lung cancer. Int J Radiat Oncol Biol Phys 77:867–876
87. Cai XW, Shedden KA, Yuan SH et al (2011) Baseline plasma proteomic analysis to identify biomarkers that predict radiation-induced lung toxicity in patients receiving radiation for non-small cell lung cancer. J Thorac Oncol 6:1073–1078
88. Chung SI, Seong J, Park YN et al (2010) Identification of proteins indicating radiation-induced hepatic toxicity in cirrhotic rats. J Radiat Res (Tokyo) 51:643–650
89. Menard C, Johann D, Lowenthal M et al (2006) Discovering clinical biomarkers of ionizing radiation exposure with serum proteomic analysis. Cancer Res 66:1844–1850
90. Oh JH, Craft JM, Townsend R et al (2011) A bioinformatics approach for biomarker identification in radiation-induced lung inflammation from limited proteomics data. J Proteome Res 10:1406–1415
91. Pevsner PH (2009) Biomarkers of ionizing radiation. US Patent 20,090,289,182
92. Pietrowska M, Polanska J, Walaszczyk A et al (2011) Association between plasma proteome profiles analysed by mass spectrometry, a lymphocyte-based DNA-break repair assay and radiotherapy-induced acute mucosal reaction in head and neck cancer patients. Int J Radiat Biol 87:711–719
93. Rithidech KN, Honikel L, Rieger R et al (2009) Protein-expression profiles in mouse blood-plasma following acute whole-body exposure to (137)Cs gamma rays. Int J Radiat Biol 85:432–447
94. Spencer SJ, Bonnin DA, Deasy JO et al (2009) Bioinformatics methods for learning radiation-induced lung inflammation from heterogeneous retrospective and prospective data. J Biomed Biotechnol 2009:Article ID 892863
95. Widlak P, Pietrowska M, Wojtkiewicz K et al (2011) Radiation-related changes in serum proteome profiles detected by mass spectrometry in blood of patients treated with radiotherapy due to larynx cancer. J Radiat Res (Tokyo) 52:575–581
96. Ao X, Zhao L, Davis MA et al (2009) Radiation produces differential changes in cytokine profiles in radiation lung fibrosis sensitive and resistant mice. J Hematol Oncol 2:6
97. Christensen E, Pintilie M, Evans KR et al (2009) Longitudinal cytokine expression during IMRT for prostate cancer and acute treatment toxicity. Clin Cancer Res 15:5576–5583
98. Hart JP, Broadwater G, Rabbani Z et al (2005) Cytokine profiling for prediction of symptomatic radiation-induced lung injury. Int J Radiat Oncol Biol Phys 63:1448–1454
99. Kong FM, Ao X, Wang L et al (2008) The use of blood biomarkers to predict radiation lung toxicity: a potential strategy to individualize thoracic radiation therapy. Cancer Control 15:140–150
100. Partridge MA, Chai Y, Zhou H et al (2010) High-throughput antibody-based assays to identify and quantify radiation-responsive protein biomarkers. Int J Radiat Biol 86:321–328
101. Rube CE, Palm J, Erren M et al (2008) Cytokine plasma levels: reliable predictors for radiation pneumonitis? PLoS One 3:e2898
102. Marchetti F, Coleman MA, Jones IM et al (2006) Candidate protein biodosimeters of human exposure to ionizing radiation. Int J Radiat Biol 82:605–639

103. Zhao L, Wang L, Ji W et al (2009) Elevation of plasma TGF-beta1 during radiation therapy predicts radiation-induced lung toxicity in patients with non-small-cell lung cancer: a combined analysis from Beijing and Michigan. Int J Radiat Oncol Biol Phys 74:1385–1390
104. Gabay C, Kushner I (1999) Acute-phase proteins and other systemic responses to inflammation. N Engl J Med 340:448–454
105. Goltry KL, Epperly MW, Greenberger JS (1998) Induction of serum amyloid A inflammatory response genes in irradiated bone marrow cells. Radiat Res 149:570–578
106. Muller K, Meineke V (2007) Radiation-induced alterations in cytokine production by skin cells. Exp Hematol 35:96–104
107. Satoh T, Hosokawa M (1998) The mammalian carboxylesterases: from molecules to functions. Annu Rev Pharmacol Toxicol 38:257–288
108. Lee WM, Galbraith RM (1992) The extracellular actin-scavenger system and actin toxicity. N Engl J Med 326:1335–1341
109. Chen Y, Zhu J, Lum PY et al (2008) Variations in DNA elucidate molecular networks that cause disease. Nature 452:429–435
110. Andreyev HJ, Wotherspoon A, Denham JW et al (2010) Defining pelvic-radiation disease for the survivorship era. Lancet Oncol 11:310–312
111. Andreyev HJ, Wotherspoon A, Denham JW et al (2011) "Pelvic radiation disease": new understanding and new solutions for a new disease in the era of cancer survivorship. Scand J Gastroenterol 46:389–397
112. Altshuler D, Daly MJ, Lander ES (2008) Genetic mapping in human disease. Science 322:881–888
113. Schadt EE (2009) Molecular networks as sensors and drivers of common human diseases. Nature 461:218–223
114. Chaze T, Slomianny MC, Milliat F et al (2013). Alteration of the serum N-glycome of mice locally exposed to high doses of ionizing radiation. Mol Cell Proteomics. 2012 Nov 12. [Epub ahead of print].

The Urine Proteome as a Radiation Biodosimeter

Mukut Sharma and John E. Moulder

Abstract

The global rise in terrorism has increased the risk of radiological events aimed at creating chaos and destabilization, although they may cause relatively limited number of immediate casualties. We have proposed that a self-administered test would be valuable for initial triage following terrorist use of nuclear/radiological devices. The urine proteome may be a useful source of the biomarkers required for developing such a test. We have developed and extensively used a rat model to study the acute and late effect of total body (TBI) and partial body irradiation on critical organ systems. This model has proven valuable for correlating the structural and functional effects of radiation with molecular changes. Results show that nephron segments differ with regard to their sensitivity and response to ionizing radiation. The urine proteome was analyzed using LC-MS/MS at 24 h after TBI or local kidney irradiation using a 10 Gy single dose of X rays. LC-MS/MS data were analyzed and grouped under Gene Ontology categories Cellular Localization, Molecular Function and Biological Process. We observed a decrease in urine protein/creatinine ratio that corroborated with decreased spectral counts for urinary albumin and other major serum proteins. Interestingly, TBI caused greater decline in urinary albumin than local kidney irradiation. Analysis of acute-phase response proteins and markers of acute kidney injury showed increased urinary levels of cystatin superfamily proteins and alpha-1-acid glycoprotein. Among proteases and protease inhibitors, levels of Kallikrein 1-related peptidase b24, precursor and products of chymotrypsin-like activity, were noticeably increased. Among the amino acids that are susceptible to oxidation by free radicals, oxidized histidine levels were increased following irradiation.

M. Sharma (✉)
Veterans Administration Medical Center, F2-120, Nephrology, Research Service, 4801 E. Linwood Blvd., Kansas City, MO 64128-2226, USA
e-mail: Mukut.Sharma@va.gov

J.E. Moulder
Department of Radiation Oncology, Medical College of Wisconsin, Radiation Biology, 8701 Watertown Plank Road, Milwaukee, WI 53226, USA

Our results suggest that proteomic analysis of early changes in urinary proteins will identify biomarkers for developing a self-administered test for radiation biodosimetry.

> **Keywords**
> Proteome • Plasma • Plasma proteins • Ionizing radiation • Radiological terrorism • Kidney • Renal • Dosimetry • Biomarkers • Triage • Proteinases • Proteinase inhibitors

5.1 The Risk of Exposure to Ionizing Radiation: Past, Present and Future

Geo-political issues and international relations continue to be decided by a struggle for possession of nuclear weapons by nation states and by terrorist organizations. Nevertheless, the USA and the Russian Republic have gradually reduced their nuclear stockpiles through mutually-binding treaties addressing nuclear proliferation. Reduction in the number of nuclear armaments controlled by powerful nations does not guarantee a world free of nuclear warfare, but it has resulted in a reduced concern over the likelihood of nuclear warfare between these countries.

Dismantling of nuclear weapon systems has increased the likelihood of smaller quantities of radioactive material reaching terrorist organizations. Such extremists may detonate small-scale improvised nuclear devices or use radiation dispersal devices (dirty bombs) in densely populated areas. In contrast to mass casualties caused by nuclear weapons, terrorist-initiated detonation of radiation dispersal devices is likely to kill smaller number of people. However, such events are intended to disrupt life by generating fear, chaos and confusion along with injuries and property damage [1].

Radioactive materials have found ever increasing peacetime applications that range from outer space exploration to geophysical discoveries and from medicine to industrial manufacturing. Nuclear energy has become a competitive source of electricity globally on land as well as under water to operate submarines, and according to the International Atomic Energy Association, up to 75% of energy requirements are met by nuclear power in some countries. Nuclear power plants date back to the early 1950s, and many of these plants will require upgrading for future use or will have to be shut down. Poor maintenance or inefficient containment during shut-down will be a risk for radiation leaks. While most radioactive materials used in industry and medicine are highly regulated, these devices are located at in many separate facilities.

The risk of radiological events has shifted from mass destruction by nuclear weapons to smaller incidents caused by terrorist-operated devices or accidental release of radioactive materials. Indeed, there have been no war time fatalities by the use of nuclear weapons since 1945; almost all cases of radiation injury since have been due to accidents or natural disasters. Recent damage to the Fukushima Daiichi nuclear power plant in Japan following an earthquake and a tsunami is an example of present and future risk posed by use of radioactive materials [2]. Such an event in an ill-prepared and densely populated areas would be disastrous. Increased 'peace time' use of radioactive materials has increased the likelihood of exposing smaller cohorts of populations to low, but biologically significant, doses of radiation [3].

5.2 Demography of the Victims of Radiological Events in Future

Terrorist actions will aim to paralyze life in the target region through chaos, fear and property damage. These events are postulated to result in a limited number of immediate deaths, but in a larger number of casualties through a combination of radiation, fire and falling/flying debris. Accidental release of radioactive material into atmosphere at nuclear power plants, as evidenced by historical and recent disasters, will likely spread depending upon natural factors such as water and wind [3]. Thus, the risk of exposure to ionizing radiation has moved from a doomsday scenario to limited casualties accompanied by destabilization of normal order.

It's likely that a radiological event will cause breakdown of communication and transportation services leaving people isolated and without assistance and these will add to the perceived danger and chaos in the affected regions. It's unlikely that a large numbers of medical facilities can be prepared for such disasters even with unrestricted access to healthcare. Further, such radiological events combined with natural disasters would increase the burden on emergency medical services. Damage to healthcare facilities by the disaster in the area may render these facilities inaccessible or non-functional. Transformation in the nature of the risk, and the demography of the affected populations, has shifted the paradigm for managing the aftermath of a radiological event. A small number of victims, directly injured by fire, debris or impact or a combination of any of these, will require assistance for mobility, communication and care. In addition to the amount of radioactive material released, the extent of radiation dose to the victims will be determined by factors such as weather conditions and building materials.

As in other emergencies, prompt screening and triage would help identify those in need of immediate medical attention and treatment. However, it's likely that most burden on civil and healthcare services will be imposed by those least affected by low-dose radiological events. Availability of a low-tech, convenient and self-administered test may be valuable in reducing the burden on public services and healthcare providers. Prior training in using such a device can enable the survivors to perform preliminary biodosimetry that would help alleviate fear and minimize chaos without overwhelming the providers of first-line emergency care [4].

Our approach to radiation dosimetry and the search for biomarkers originates from the need for simple self-administered assessment of radiation biodosimetry. We postulated that identification of biomarkers using noninvasively obtained biomaterials such as the urine or saliva would lead to the development a device for self administered test. The following discussion summarizes our efforts to identify biomarkers. Initial work based on the urine proteome has provided encouraging results.

5.3 A Rat Model to Study Radiation Dosimetry

A consistent and reliable source of biomaterial is a prerequisite for efforts to identify biomarkers of radiation injury. Radiation exposure studies using human volunteers cannot be planned for obvious ethical reasons, and relevant human specimens could be obtained only after a radiological event. Thus preliminary studies using rodents are needed to provide the information needed to plan studies using higher species. Therefore, we developed a rat model to study the late effects of total body irradiation (TBI) that results in organ failure and mortality. Earlier studied were focused on late effects of ionizing radiation on renal function. However, recent studies have provided valuable information to understand the effect of TBI on several tissues.

If WAG/RijCMCR rats are given 10 Gy TBI and hematopoietic toxicity is prevented by a bone marrow transplant, proteinruria develops about 6 weeks later, followed shortly by azotemia and hypertension [5]. At this TBI dose, azotemia exceeds 120 mg/dl by about 20 weeks and animals must be euthanized [6]. The renal effects of TBI in this mdoel can be mitigated by post-irradiation therapy with either angiotensin converting enzyme inhibitors or angiotensin II receptor blockers [6]. Although radiation nephropathy is the principal cause of morbidity in this TBI model, cardiac [7] and hepatic [8] injury can also be observed. At higher radiation doses, radiation pneumonitis can be a cause or morbidity prior to the development of radiation nephrapathy [9].

5.4 Renal Structure and Function Studies Using the Rat Model

Initial studies included experiments to determine whether early effects of X-irradiation cause demonstrable changes in the kidney nephron. We used an in vitro assay of glomerular barrier function as a surrogate for increased passage of plasma proteins. Previously we had used this assay to demonstrate that increased glomerular albumin permeability precedes the onset of proteinuria in models of human disease [10–14]. We demonstrated increased glomerular albumin permeability at 1 h after irradiation in the rat TBI model. Initial studies also showed that 10Gy X-irradiation did not alter glomerular protein permeability barrier characteristics in C57BL6 mice [15]. Thus, our rat model provides a suitable rodent model to study effects of ionizing radiation on renal structure, function and urinary proteome.

We postulated that altered glomerular permeability to macromolecules indicates early renal injury that precedes overt proteinuria. Recent work showed that immediate increase in glomerular permeability to macromolecules following TBI is not paralleled by changes in glomerular ultrastructure [16]. The observed latency between early glomerular change and the onset of proteinuria and histological changes may be a valuable window to study the processes that cause systemic and renal changes leading to proteinuria. These observations also suggested that subtle renal injury caused by TBI may alter the urinary protein profile in ways not detectable by conventional urinalysis. These qualitative changes in urinary protein composition will require sensitive techniques for detection and identification of molecules.

is the primary structural and functional unit in the kidney. Structurally connected glomerulus, proximal tubule, Loop of Henle and distal tubule leading to the collecting duct constitute each nephron. These microscopic structures function in concert to remove metabolic waste from the circulating blood. The plasma filtrate generated in the glomerulus undergoes several modifications in each tubular segment of the nephron prior to its conversion into urine and excretion. During its passage through the tubule most of the proteins/peptides, hormones and other metabolites are reabsorbed, mono and divalent ions are reabsorbed and fluid volume adjusted. These functions are performed by distinct cell types in different sections of the tubules in each nephron.

Immediate changes in glomerular permeability characteristics, as noted above, without increased urinary protein suggest differences in the effect of radiation on the glomerular and tubular sections of the nephron. Follow up studies showed that tubular injury does not accompany increased glomerular albumin permeability. We studied endocytosis of bovine serum albumin using isolated tubular preparations from irradiated rats. These experiments suggested that proximal tubular capacity to reabsorb albumin from the ultrafiltrate is not altered immediately after TBI. Instead, the tubular capacity to reuptake albumin decreases several weeks after TBI corresponding to the time of the onset of proteinuria [17]. Thus, within the kidney, glomeruli, tubules, interstitium and the vasculature differ with regard to sensitivity and/or response to ionizing radiation [15, 18]. These observations were valuable in interpreting the recent results using proteomic technology as described below. These observations also emphasize the importance of consistently employing the same animal model to study radiation injury.

5.5 Variations in Tissue Response to Radiation in the Rat Model

Our rat model provides a convenient opportunity to determine differences within structural components of the functional organ. Nephron

5.6 Partial Body Irradiation Versus Non-homogeneous Irradiation

In addition to variations in tissue sensitivity to radiation the tissue volume exposed and the dose of radiation absorbed by the tissues also determine the long term effects of ionizing radiation. Exposure to a radioactive source by shielding the surroundings may results in partial-body irradiation. A better understanding of the effects of total body irradiation (TBI) or partial body or non-homogeneous irradiation is important in countermeasure strategies. Thus we considered a scenario in which radiation is not delivered or absorbed uniformly, and conducted studies using both TBI and localized kidney irradiation [19].

5.7 Methods for Radiation Biodosimetry and Biomarker Discovery

Multiple methods for radiation biodosimetry are in development. These methods range from physical examination to molecular analysis. An impressive amount of work has been accomplished in several specialties during the last 10 years, but it is difficult to identify a single technology of choice at the present time. It is possible that a systematic approach to radiation biodosimetry will be based on a combination of methods. We have used LC-MS/MS to identify proteins in the urine from irradiated rats [16]. Parallel work in our laboratory has focused on the human plasma proteome to study chronic kidney disease called focal segmental glomerulosclerosis [20]. Detailed analysis of the urine proteome may lead to biomarker(s) suitable for developing a self-administered test for early triage in the victim of a radiological event.

There is considerable interest in using early clinical examination of the victims for triage decisions. Gastrointestinal effects (i.e., emesis or severe diarrhea) have been evaluated for radiation biodosimetry [21, 22], but this may not actually very useful in practice [23]. Radiation injury may cause a cutaneous syndrome and associated changes in skin blood flow, cytology and cytokines, but the delayed appearance (7–14 days after exposure) of changes on the skin surface limits the usefulness of this clinical parameter for biodosimetry.

Lymphocyte cytology to determine the number of bicentric cells is currently considered the gold-standard in radiation dosimetry [24, 25]. Analysis of frequency distribution of chromosomal aberrations provides in depth determination of the injury. Ongoing efforts to automate this sophisticated technique may make it less labor intensive and improve its throughput [26].

Biophysical techniques such as electron paramagnetic resonance (EPR) spectroscopy and optically stimulated luminescence (OSL) are potentially useful tools for developing biodosimetry algorithms. These techniques, once fully developed, may be suitable for determining partial body irradiation in the extremities or head and neck areas. EPR (electron spin resonance, ESR) spectroscopy can detect stable radicals in tooth enamel in vivo as well as in and nail clippings ex vivo [27–29]. However, OSL technology is in the early stages of development for retrospective dosimetry [30].

Molecular techniques including gene expression analysis, metabolomics and proteomic may provide details on molecular phenomena associated with total body or organ-specific effects of radiation. Several studies using peripheral lymphocyte have described temporal profiles of gene expression after various doses of radiation [31–33]. Recent studies indicate that expression of tissue specific RNA species such as micro-RNAs may provide molecular tools to establish radiation dosimetry after total or partial body irradiation [34–36].

Metabolomic and proteomic analyses provide survey of the products of changes in gene expression. Both nuclear magnetic resonance (NMR) spectroscopy and mass spectrometry (MS) technologies have been tested in metabolomics studies. These techniques offer sensitive and high throughput analysis of biological samples to identify low molecular weight biomarkers. These techniques differ with regard to sensitivity, sample preparations or experimental design, but they have provided partially comparable results. Most noticeable molecules identified in these studies belong to energy metabolism, amine metabolism, lipid metabolism and nucleotide metabolism [37–39].

5.8 Proteomics Technology for Biomarker Discovery

Mass spectrometry (MS) is an integral part of proteomic technology for high throughput and sensitive analysis of biological materials. Besides its application in metabolomic studies to identify simple organic molecules, MS has been increasingly used to detect and identify proteins ranging from small peptides to complex proteins. Consistent development of methodology has made MS an indispensable technology to interpret the functional aspect of genomic information through a vast array of analytical strategies. Urine proteomics techniques continue to evolve for in depth understanding of disease related changes in urinary proteins [40, 41].

Sample preparation and separation are critical components; and recommended methods for collection, storage, initial treatments of urine samples and fractionation are available [42–44]. Proteomic methodologies for protein identification in samples fall under two complementary categories namely, bottom-up and top-down methodologies. In bottom-up proteomics approaches samples are first treated with a protease to obtain small peptides for ionization. Analysis of proteins in the tissue/biofluid can be carried out using two approaches. Protein fragments can be analyzed as a mixture or after separation and fractionation using one/two dimensional (1-DE/2-DE) electrophoresis or liquid chromatography (LC). Protein fragments are analyzed using tandem mass spectrometry where a peptide ion is detected first (MS analysis) followed by next level fragmentation using collision-induced dissociation (CID) to peptide product ions (MS/MS analysis). Analytical results are obtained as the mass/charge (m/z) ratio of each charged molecular species. The peptide-mass fingerprint consisting of information regarding each peptide is unique to a specific protein. Observed masses can be compared with theoretical in silico sequences to identify proteins. Thus, bottom-up approaches involve a two-stage protein fragmentation to generate ionic species followed by protein identification. However, secondary features that define the molecular role, pos-translational modifications, are not included in these analyses.

Top-down proteomics approaches involve direct MS analysis of intact proteins without proteolysis. Intact proteins in biomaterials are separated by 1-DE/2-DE or LC and analyzed by matrix-assisted laser desorption/ionization (MALDI)-MS or Electrospray ionization mass spectrometry (ESI-MS). MS/MS fragmentation of protein ions is achieved by Electron Capture Dissociation (ECD) or Electron Transfer Dissociation (ETD) in ESI-MS ionization method or by in-source decay (ISD) in MALDI-TOF analysis. Another top-down approach involves Fourier-transform-ion cyclotron resonance (FT-ICR) MS with ECD or ETD for protein ion fragmentation. Top-down analyses result in extremely complex spectra, but the results include post-translational modifications and C- and N-terminal sequences that are not obtained using the top-down approach [45, 46].

5.9 Proteomic Data Analysis

Mass spectrometry results are analyzed using algorithms established for the equipment by both the manufacturer and the investigator. During recent studies we searched MS/MS spectra against the rodent portion of the UniProt database using a two-stage process. First, MS/MS spectra peak lists (.*dta* files) were extracted from the .RAW data files generated by the MS system based on the *extract_ms* program (ThermoFisher, Waltham, MA). The following parameters served as the basis for searches: B 600, T 3500, M 1.4, S1 G1, E 250. The MS/MS data was searched against the rodent subset of the UniProt database (release 54.0) with a version of the *Sequest* (V.12) program adapted for use on the JS-20 processor (IBM, Armonk, NY) in a cluster environment (ThermoFisher, Waltham, MA). The following key parameters were employed. For the first pass search: Enzyme Trypsin [KR], peptide tolerance 2.5 AMU, fragment ions tolerance 1.0 AMU, differential modifications: M 16, C 57. For the second pass search: Enzyme No enzyme, peptide tolerance 2.5 AMU, fragment ions tolerance 1.0

AMU, differential modifications: MHP 16, C 57, K 1, R -47. Next, the search results (.*out files*) were parsed and the peptide and protein probabilities calculated using the *Epitomize* program developed at the Medical College of Wisconsin (MCW, Milwaukee, WI) and the protein hits with probability scores >0.6 from each search were saved. The protein hits from the individual subfractions derived from the original pooled urine samples were combined using the *Visualize* program developed at the MCW and revised protein probability scores calculated based on the observation of proteins in multiple runs, and proteins with a probability score of 0.90 retained [47]. By searching similar data with decoy databases, it was found that this probability score corresponds to a False Discovery Rate (FDR) of 5% and 7.5% in separate studies [48]. These composite results were then quantitatively compared using the *Visualize* program and the ratios of protein abundance were exported for further analysis by *Excel* (Microsoft, Redmond, WA).

Proteins were grouped using Gene Ontology (GO) categories: Cellular Localization/Component, Molecular Function and Biological Process [49].To better understand the changes in the global patterns of urine protein abundance caused by irradiation, proteins were divided into three groups: increased twofold or greater in irradiated urine, decreased less than twofold in irradiated urine, or remained unaffected unchanged. Each of the proteins in each of the samples was mapped to the GO.

Visualize program was used for quantitative GO analysis for the categories Cellular Localization, Molecular Function and Biological Process [50]. In order to avoid multiple counting of spectra that can be assigned to more than one protein, we adopted the following strategy. For each spectrum, the GO terms for every protein that could have generated the peptide that best matched the spectrum were collected. Each GO term was assigned a fractional value based on the number of proteins assigned to the spectrum and the total number of GO terms assigned to each of the proteins for each of the three GO categories. Thus, for each spectrum, the sum of all GO terms from all proteins equaled one for each of the three GO categories. Expressing proteins as fractional scans minimizes repeat count, and provides a more accurate estimate of protein abundance than a simple scan count.

MS technology is applicable for exploring the entire proteome in a single analysis. Quantification of MS results has become feasible through strategies that involve incorporating a stable isotope into cells either during culture or by comparing signals from a peptide under control and experimental conditions. Direct quantification of MS data can be facilitated through introduction of a predictable mass difference through post-harvest incorporation of stable isotopes into peptides during trypsin digestion. Identity of candidate molecules can be further confirmed using specific antibodies or antibody arrays [51–53].

5.10 Outline of Experimental Approach

We have completed specific initial studies to determine the effect of TBI versus local kidney irradiation using a single 10 Gy dose of X-ray at 24 h post-irradiation. Rats used in these studies were 8 weeks-old male WAG/RijCMCRs that were bred and housed in a moderate-security barrier at the MCW. Animals were maintained in the MCW Biomedical Resource. Conscious animals were held in plastic restrainers and administered 10 Gy orthovoltage X-rays in one. Local kidney irradiation was done with parallel-opposed lateral fields; the field size was large enough to account for the uneven positioning of the kidneys. Control rats were sham irradiated. Previously published dosimetry protocols were followed [54]. Rats were placed in metabolic cages and 24 h urine samples were collected. Urine samples were centrifuged (3,000 g) to remove particulate contamination and aliquots were used for total protein, creatinine assays or proteomic analysis. Samples were prepared for LC-MS/MS following protocols referred to earlier.

5.11 Summary of the Results and Their Implications for Future Work to Identify Biomarkers

Our approach to biomarker discovery has included analysis of structural and functional changes in the nephron along with proteomic analysis of urine. These preliminary studies were planned to obtain information on the effect of ionizing radiation on glomerular ultra-structure and function, urinary protein and creatinine and urine proteome using LC-MS/MS. Results of these studies were presented in recent publications. Detailed results of mass spectrometry based identification of urine proteins are included in these articles [16, 55]. The following discussion is focused on some of the key findings that we consider valuable for future work on biomarker identification and for developing self-administered test for convenient and rapid triage.

5.11.1 Effect of Ionizing Radiation on Nephron Function

Our results show that glomerular function is altered almost immediately after TBI (10 Gy) as indicated by increased permeability to albumin in vitro [15, 16, 56]. These findings are consistent with parallel studies on the effect of other agents such as free radicals, cytokines (TGF-β and TNFα), antibodies to glomerular proteins or human plasma factor that cause increased glomerular protein as mentioned earlier. However, irradiated rats develop proteinuria only about 6 weeks after TBI or local kidney irradiation. These observations are in line with other studies where increased glomerular albumin permeability precedes development of proteinuria. Rat models of puromycin aminonucleoside nephrosis, type 2 diabetes and radiation nephropathy show increased glomerular permeability prior to the onset of albuminuria [10, 12, 16]. These observations on early increase in glomerular permeability and late development of proteinuria are reconciled by our observations on the tubular effects of radiation. These experiments show that despite increased glomerular albumin permeability, proximal tubular reabsorption of albumin is maintained for weeks after TBI [17]. Similar observations were made in the puromycin aminonucleoside-induced nephrosis model of nephrotic syndrome. In this model of minimal change disease a single intraperitoneal injection of puromycin aminonucleoside causes immediate increase in glomerular albumin permeability with detectable proteinuria only after several days [10]. These observations indicate that effects of ionizing radiation are cell-type specific and that functional changes leading to organ dysfunction are progressive. Therefore, detection of injury and strategies for mitigation and treatment should include these variables. While complex, progressive nature of injury also provides the opportunity for targeted intervention.

5.11.2 Effect of Ionizing Radiation on Clinically Used Parameter Urine Protein/Urine Creatinine (Up/Uc) Ratio

The Up/Uc ratio is considered an important clinical milestone in the development of chronic disease [57]. We compared the effect of TBI and local kidney irradiation on Up/Uc ratio as an indicator of renal function. Both TBI and local kidney irradiation resulted in short-term decrease in the Up/Uc ratio [16, 55]. Decreased urinary protein indicates a quantitative reduction in the amount of total protein that may reflect vascular and neurohumoral changes that result in altered renal blood flow [18]. Another reason for a decrease in urinary protein is a likely increase in tubular reabsorption of certain proteins specially macromolecules e.g., albumin. Indeed, two-photon intravital microscopy showed increased tubular reabsorption of filtered albumin immediately in the early phase after puromycin aminonucleoside (PAN) treatment [10]. Increase in the number of proteins detected by silver-staining of 2DE gels with a quantitative decrease in the urinary protein is an intriguing observation that's being actively studied by our group.

Up/Uc is a candidate for an initial tool kit to identify early markers of radiation injury because it also provides a method to differentiate the early phase of radiation injury from other chronic diseases including pre-existing conditions. An important consideration in selecting a biomarker is the specificity of the candidate indicators to the pathophysiological condition. Increased urinary protein is an independent risk factor of cardiovascular disease. Increased Up/Uc is associated with systemic diseases including diabetes and hypertension. In our model, rats develop overt proteinuria with increased Up/Uc about 6 weeks after TBI. Thus, a decreased Up/Uc ratio immediately after a radiological event would indicate sudden change in renal protein handling in healthy individuals as well as in those with pre-existing pathophysiological conditions. Additional studies will be needed to validate these ideas under various experimental conditions, but we think this classical indicator of renal function may provide useful clues to early signs of injury caused by irradiation.

5.11.3 Ionizing Radiation and Specific Groups of Proteins in the Urine Proteome

As expected, proteomic analysis of urine proteins provided a large amount of data. Here we discuss some of the observations that may prove to be potential leads for future work.

5.11.3.1 Albumin and Other Major Plasma Proteins

In line with the observed decrease in Up/Uc ratio, proteomic data showed decreased urinary albumin and other major plasma proteins after both TBI and local kidney irradiation. However, a comparison of the effect of TBI and local kidney irradiation on urinary albumin provides intriguing results. Proteomic spectral analysis shows that TBI resulted in a three- to fourfold greater decline in urinary albumin compared to local kidney irradiation [16, 55]. Immunoglobulins, another important class of plasma proteins, also showed decreased presence in the urine post-irradiation. Since protein filtration and reabsorption are renal events, plasma and other tissues would be expected to remain unaltered by local kidney irradiation. However, the relatively greater effect of TBI compared to local kidney irradiation suggests a role of extra-renal factors in the renal handling of plasma proteins.

5.11.3.2 Acute Phase Proteins

We attempt to understand the early effects of irradiation in terms of acute kidney injury (AKI). Acute kidney injury may increase the risk for chronic kidney disease and end-stage renal disease. AKI is now considered as an independent risk factor for chronic kidney disease, end-stage renal disease, other renal complications and death [58]. A number of new candidate markers of AKI have been proposed and a few are now recommended for clinical use. We analyzed our LC-MS/MS data to identify markers of AKI, Kidney Injury Molecule-1 (KIM), N-acetyl-Beta-D-glucosaminidase, Neutrophil gelatinase-associated lipocalin (NGAL), Cystatin C and Interleukin-18 (IL-18) [59, 60]. In one study we found that TBI caused a threefold increase in the levels of Cystatin C compared to the control. However, KIM, neutrophil gelatinase, N-acetyl beta-D-glucosaminidase or IL-18 were not detected in the urine after TBI or local kidney irradiation [16, 55].

Members of the cystatin superfamily include fetuins, cystatin-C and histidine rich glycoprotein. We found that TBI (10 Gy) resulted in increased urinary fetuin-B in one study [16]. We also found that 2, 5 or 10 Gy TBI without BMT also increased the MS spectral count in the urine up to 5 days [unpublished observations]. Fetuin and cystatin-C are also considered biomarkers of AKI [61]. Additionally, Alpha-1-acid glycoprotein (orosomucoid) is a circulating acute phase protein that is known for its transport function. Increased urinary alpha-1-acid glycoprotein has previously been associated with trauma, stress and type II diabetes [62, 63]. These observations suggest that while radiation injury results in excretion of some of the proteins associated with AKI (cystatin-C), majority of the recently identified markers of AKI are not affected. Thus, radiation injury and AKI may differ with regard to their mechanisms and biomarkers.

5.11.3.3 Proteinases and Proteinase Inhibitors

We used Gene Ontology (GO) analysis to identify proteins under GO categories of cellular localization/component, molecular function, and biological process [49]. These analyses showed increased levels of proteinases in the urine proteome. These proteinases represent both plasma and cellular origins. The most noticeable increase occurred in the levels of kallikrein 1-related peptidase b24 precursor (Glandular kallikrein K24, Tissue kallikrein 24, gi|48428341). We also found that products of chymotrypsin-like activity are detectable in the 24 h urine after TBI [16]. Increased numbers of low abundance proteinases may partly explain the observed increase in protein counts without a parallel increase in the quantities of urine protein determined by spectrophotometric methods.

Proteinases and corresponding proteinase inhibitors are important in cellular response to injury and several classes of these molecules are present in the circulation and renal cells. Several proteinases are critical for tubular function in addition to their role in the complement cascade. A parallel decrease in several proteinase inhibitors with the observed increase in urinary proteinases was noticeable. These protinase inhibitors included kallikrein binding serine protease inhibitor A3K precursor (SERPIN A3K, gi|266407). Changes in kallikrein-kinin system proteins may influence vascular function, coagulation and inflammation processes. Plasma and tissue kallikreins act on precursor kininogens to synthesize vasodilatory bradykinin or kallidin peptides. Carboxypeptidases, neutral endopeptidases and angiotensin converting enzymes metabolize and inactivate bradykinin and kallidin. Increased urinary kallikrein and inhibitor are biologically significant and may be valuable in understanding the mechanism of development of radiation injury [64, 65].

5.11.3.4 Amino Acid Oxidation

Ionizing radiation causes oxidative damage and chemical changes in DNA, lipids and proteins. We speculated that TBI-induced injury may result in excretion of modified proteins in the urine [66, 67]. MS/MS spectra were queried using protein databases for amino acid oxidation as a dynamic modification through reaction with the reactive oxygen species. Superoxide-induced modifications of amino acids include the oxidation of methionine (M), histidine (H), lysine (K) and proline (P) as well as the formation of glutamic acid semialdehyde from arginine (R). Results showed that TBI induced a decrease in oxidized methionine and a nearly threefold increase in oxidized histidine after TBI [16].

Selective oxidation of histidine during the early stages may have significant implications since histidine is an essential amino acid for humans and rats. Histidine deficiency alters protein turnover rate and levels of hemoglobin, albumin and transferrin [68]. Histidine contains an imidazole ring (neutral pKa) as a side chain that permits its participation in acid/base catalysis and proton transfer reactions at physiological pH. The nucleophilicity of imidazole also makes it a good ligand for electrophilic metal ions, such as zinc in many enzymes e.g., carboxypeptidase. Histidine is also a key amino acid in the hemoglobin molecule since one histidine residue stabilizes the porphyrin structure through a covalent linkage with the heme iron. Another histidine molecule blocks oxidation of iron, prevents carbon monoxide binding with heme iron while allowing oxygen to bind easily. Oxidative damage to protein histidine by hydroxyl radicals or products of lipid oxidation such as HNE (4-hydroxynon-2-enal) has been reported [69]. Radiation-induced increase in oxidized histidine may affect acid/base balance by enzymes such as carbonic anhydrase and oxygen transport by hemoglobin.

5.12 Conclusions

Soon after the 2001 attacks on the World Trade Center and the Pentagon, scientists began to more seriously assess our ability to cope with a radiological terrorism incident [70]. The outline of what was needed was fairly obvious: the ability to prevent an attack; methods to cope with the medical consequences; the ability to clean up afterwards; and the tools to figure out who did it [71]. In dealing with the medical consequences there were three major issues: determining exposure, dealing with acute radiation injuries, and dealing with chronic radiation injuries [21, 72]. Considerable effort is being made to improve treatment of acute radiation injuries [73–75]; and in the laboratory, a wide range of late normal tissue injuries can be mitigated using post-irradiation therapies [6, 9, 76–79].

In practice, medical intervention requires knowing radiation doses, preferably organ-specific doses [19, 21, 80]. While there are widely-available instruments for assessing contamination, the tools for retroactive assessment of radiation dose are either primitive or not widely available [21, 22]. If a mass-casualty incident occurred now, the only method for rapid (less than 12 h) dose assessment would be "time to emesis" [21, 22], but this may not actually very useful in practice [23]. If more time were available, dose estimates could be made based on lymphocyte depletion kinetics, but that takes at least a day [21, 22]. In theory, doses could also be based on chromosome aberrations in blood lymphocytes, but this assay takes days [21].

Urine proteomics holds promise for providing triage biodosimetry and possibly also for dose determination and evaluation of doses to specific organs. Our studies have provided preliminary data and groups of proteins that will be useful for further investigations [16, 55]. Of course, much more needs to be done. We need to understand the time course of the changes and their dependence on radiation dose. Perhaps more importantly, we need to determine which changes are radiation specific, as in most mass casualty scenarios there will be other causes of injury including trauma and burns. If the optimal urine proteins can be identified, there is the potential to develop biodosimetry based on robust, but simple-to-use, technologies such as a urine dip sticks.

References

1. National Research Council (2009) Assessing medical preparedness to respond to a terrorist nuclear event: workshop report. The National Academy Press, Washington, DC
2. Sarin R (2011) Chernobyl, Fukushima, and beyond: a health safety perspective. J Cancer Res Ther 7(2):109–111
3. National Council on Radiation Protection and Measurements (2001) Management of terrorist events involving radioactive material. NCRP report no. 138, Bethesda
4. Swartz HM, Flood AB, Gougelet RM, Rea ME, Nicolalde RJ, Williams BB (2010) A critical assessment of biodosimetry methods for large-scale incidents. Health Phys 98(2):95–108
5. Lenarczyk M, Cohen EP, Fish BL, Irving AA, Sharma M, Driscoll CD et al (2009) Chronic oxidative stress as a mechanism for radiation nephropathy. Radiat Res 171(2):164–172
6. Moulder JE, Cohen EP, Fish BL (2011) Captopril and losartan for mitigation of renal injury caused by single-dose total body irradiation. Radiat Res 175(1):29–36
7. Baker JE, Fish BL, Su J, Haworth ST, Strande JL, Komorowski RA et al (2009) 10 Gy total body irradiation increases risk of coronary sclerosis, degeneration of heart structure and function in a rat model. Int J Radiat Biol 85(12):1089–1100
8. Moulder JE, Fish BL, Holcenberg JS, Sun GX (1990) Hepatic function and drug pharmacokinetics after total body irradiation plus bone marrow transplant. Int J Radiat Oncol Biol Phys 19:1389–1396
9. Rosenthal RA, Fish B, Hill RP, Huffman KD, Lazarova Z, Mahmoud J et al (2011) Salen Mn complexes mitigate radiation injury in normal tissues. Anti-Cancer Agents Med Chem 11(4):359–371
10. McCarthy ET, Sharma R, Sharma M (2005) Protective effect of 20-hydroxy-eicosatetraenoic acid (20-HETE) on glomerular protein permeability barrier. Kidney Int 67(1):152–156
11. Dahly-Vernon AJ, Sharma M, McCarthy ET, Savin VJ, Ledbetter SR, Roman RJ (2005) Transforming growth factor-β, 20-HETE interaction, and glomerular injury in Dahl Salt-Sensitive rats. Hypetension 45:1–6
12. Sharma R, Sharma M, Reddy S, Savin VJ, Nagaria AM, Wiegmann TB (2006) Chronically increased intrarenal angiotensin II causes nephropathy in an animal model of type 2 diabetes. Front Biosci 1(11):968–976
13. Sharma M, Sharma R, Reddy SR, McCarthy ET, Savin VJ (2002) Proteinuria after injection of human focal segmental glomerulosclerosis factor. Transplantation 73:366–372

14. Sharma M, Sharma R, McCarthy ET, Savin VJ (2004) The FSGS permeability factor: biochemical characteristics and biological effects. Exp Biol Med 229:85–98
15. Sharma M, Sharma R, Ge XL, Fish BL, McCarthy ET, Savin VJ, Cohen EP, Moulder JE (2001) Early detection of radiation-induced glomerular injury by albumin permeability assay. Radiat Res 155:474–480
16. Sharma M, Halligan BD, Wakim BT, Savin VJ, Cohen EP, Moulder JE (2008) The urine proteome as a biomarker of radiation injury. Proteomics Clin Appl 2(7):1065–1086
17. Yammani RR, Sharma M, Seetharam S, Moulder JE, Seetharam B (2002) Loss of albumin and megalin binding to renal cubilin results in albuminuria in rats exposed to total body irradiation. Am J Physiol Regul Integr Comp Physiol 283(2):R339–R346
18. Rose BD (1987) Pathophysiology of renal disease. McGraw-Hill Professional, New York, pp 366–368
19. Prasanna PGS, Blakely WF, Bertho JM, Chute JP, Cohen EP, Goans RE et al (2010) Synopsis of partial-body radiation diagnostic biomarkers and medical management of radiation injury workshop. Radiat Res 173(2):245–253
20. Savin VJ, McCarthy ET, Sharma M (2012) Permeability factors in nephrotic syndrome and focal segmental glomerulosclerosis. Kidney Res Clin Pract 31:205–213
21. Rojas-Palma C, Liland A, Jerstad AN, Etherington G, del Rosario Pérez M, Rahola T et al (2009) TMT HANDBOOK—triage, monitoring and treatment—handbook for management of the public in the event of malevolent use of radiation. Norwegian Radiation Protection Agency, Østerås
22. Swartz HM, Williams BB, Nicolade RJ, Demidenko E, Flood AB (2011) Overview of biodosimetry for management of unplanned exposures to ionizing radiation. Radiat Meas 46:742–748
23. Demidenko E, Williams BB, Swartz HM (2009) Radiation dose prediction using data on time to emesis in the case of nuclear terrorism. Radiat Res 171(3):310–319
24. IAEA (2001) Cytogenetic analysis for radiation dose assessment: a manual. Technical report 405, IAEA, Vienna
25. ISO (2004) Radiation protection—performance criteria for service laboratories performing biological dosimetry by cytogenetics. ISO 19238:2004, International Organization for Standardization, Geneva
26. Garty G, Karam A, Brenner DJ (2011) Infrastructure to support ultra high throughput biodosimetry screening after a radiological event. Int J Radiat Biol 87:754–765
27. IAEA (2002) Use of electron paramagnetic resonance dosimetry with tooth enamel for retrospective dose assessment. TECDOC-1331, IAEA, Vienna
28. Swartz HM, Burke G, Coey M, Demidenko E, Dong R, Grinberg O, Hilton J, Iwasaki A, Lesniewski P, Schauer DA (2007) In vivo EPR for dosimetry. Radiat Meas 42:1075–1084
29. Trompier F, Kornak L, Calas C, Romanyukha A, Leblanc B, Mitchell CA, Swartz HM, Clairand I (2007) Protocol for emergency EPR dosimetry in fingernails. Radiat Meas 42:1085–1088
30. Godfrey-Smith DI, Pass B (1997) A new method for retrospective radiation dosimetry: optically stimulated luminescence in dental enamel. Health Phys 72:744–749
31. Blakely WF, Prasanna PG, Grace MB, Miller AC (2001) Radiation exposure assessment using cytological and molecular biomarkers. Radiat Prot Dosimetry 97(1):17–23
32. Meadows SK, Dressman HK, Muramoto GG, Himburg H, Salter A, Wei Z, Ginsburg G, Chao NJ, Nevins JR, Chute JP (2008) Gene expression signatures of radiation response are specific, durable and accurate in mice and humans. PLoS One 3:e1912
33. Kabacik S, Mackay A, Tamber N, Manning G, Finnon P, Paillier F, Ashworth A, Bouffler S, Badie C (2011) Gene expression following ionising radiation: identification of biomarkers for dose estimation and prediction of individual response. Int J Radiat Biol 87(2):115–129
34. Lewis BP, Burge CB, Bartel DP (2005) Conserved seed pairing, often flanked by adenosines, indicates that thousands of human genes are microRNA targets. Cell 120:15–20
35. Marsit CJ, Eddy K, Kelsey KT (2006) MicroRNA responses to cellular stress. Cancer Res 66:10843–10848
36. Cha HJ, Seong KM, Bae S, Jung JH, Kim CS, Yang KH, Jin YW, An S (2009) Identification of specific microRNAs responding to low and high dose gamma-irradiation in the human lymphoblast line IM9. Oncol Rep 22(4):863–868
37. Coy SL, Cheema AK, Tyburski JB, Laiakis EC, Collins SP, Fornace A Jr (2011) Radiation metabolomics and its potential in biodosimetry. Int J Radiat Biol 87(8):802–823
38. Chen C, Brenner DJ, Brown TR (2011) Identification of urinary biomarkers from X-irradiated mice using NMR spectroscopy. Radiat Res 175(5):622–630
39. Johnson CH, Patterson AD, Krausz KW, Lanz C, Kang DW, Luecke H, Gonzalez FJ, Idle JR (2011) Radiation metabolomics. 4. UPLC-ESI-QTOFMS-Based metabolomics for urinary biomarker discovery in gamma-irradiated rats. Radiat Res 175(4):473–484
40. Kortz L, Helmschrodt C, Ceglarek U (2011) Fast liquid chromatography combined with mass spectrometry for the analysis of metabolites and proteins in human body fluids. Anal Bioanal Chem 399:2635–2644
41. Zerefos PG, Aivaliotis M, Baumann M, Vlahou A (2012) Analysis of the urine proteome via a combination of multi-dimensional approaches. Proteomics—European Consortium. doi:10.1002/pmic.201100212
42. Thongboonkerd V (2007) Practical points in urinary proteomics. J Proteome Res 6(10):3881–3890

43. Apweiler R, Aslanidis C, Deufel T, Gerstner A et al (2009) Approaching clinical proteomics: current state and future fields of application in fluid proteomics. Clin Chem Lab Med 47(6):724–744
44. Court M, Selevsek N, Matondo M, Allory Y, Garin J, Masselon CD, Domon B (2011) Toward a standardized urine proteome analysis methodology. Proteomics 11(6):1160–1171
45. Mirza SP, Olivier M (2008) Methods and approaches for the comprehensive characterization and quantification of cellular proteomes using mass spectrometry. Physiol Genomics 33:3–11
46. Calligaris D, Villard C, Lafitte D (2011) Advances in top-down proteomics for disease biomarker discovery. J Proteomics 74(7):920–934
47. Halligan BD, Greene AS (2011) Visualize: a free and open source multifunction tool for proteomics data analysis. Proteomics 11(6):1058–1063
48. Wu B, Guan Z, Zhao H (2006) Parametric and nonparametric FDR estimation revisited. Biometrics 62:735–744
49. Ashburner M, Ball CA, Blake JA, Botstein D et al (2000) Gene ontology: tool for the unification of biology. The Gene Ontology Consortium. Nat Genet 25:25–29
50. Harris MA, Clark J, Ireland A, Lomax J et al (2004) The Gene Ontology (GO) database and informatics resource. Nucleic Acids Res 32(Database issue):D258–D261
51. Pan S, Aebersold R, Chen R, Rush J, Goodlett DR, McIntosh MW, Zhang J, Brentnall TA (2009) Mass spectrometry based targeted protein quantification: methods and applications. J Proteome Res 8(2):787–797
52. Brewis IA, Brennan P (2010) Proteomics technologies for the global identification and quantification of proteins. Adv Protein Chem Struct Biol 80:1–44
53. Deutsch EW, Mendoza L, Shteynberg D, Farrah T, Lam H, Tasman N, Sun Z, Nilsson E, Pratt B, Prazen B, Eng JK, Martin DB, Nesvizhskii AI, Aebersold R (2010) A guided tour of the Trans-Proteomic Pipeline. Proteomics 10(6):1150–1159
54. Cohen EP, Fish BL, Sharma M, Li XA, Moulder JE (2007) The role of the angiotensin II type-2 receptor in radiation nephropathy. Transl Res 150:106–115
55. Sharma M, Halligan BD, Wakim BT, Savin VJ et al (2010) The urine proteome for radiation biodosimetry: effect of total body versus local kidney irradiation. Health Phys 98(2):186–195
56. Sharma M, McCarthy ET, Sharma R, Fish BL et al (2006) Arachidonic acid metabolites mediate the radiation-induced increase in glomerular albumin permeability. Exp Biol Med (Maywood) 231:99–106
57. Konvalinka A, Scholey JW, Diamandis EP (2011) Searching for new biomarkers of renal diseases through proteomics. Clin Chem. doi:10.1373/clinchem.2011.165969
58. Coca SG, Singanamala S, Parikh CR (2011) Chronic kidney disease after acute kidney injury: a systematic review and meta-analysis. Kidney Int. doi:10.1038/ki.2011.379
59. Bonventre JV (2007) Diagnosis of acute kidney injury: from classic parameters to new biomarkers. Contrib Nephrol 156:213–219
60. Honore PM, Joannes-Boyau O, Boer W (2007) The early biomarker of acute kidney injury: in search of the Holy Grail. Intensive Care Med 33:1866–1868
61. Devarajan P (2008) Proteomics for the investigation of acute kidney injury. Contrib Nephrol 160:1–16
62. Kalousova M, Zima T, Tesar V, Dusilova-Sulkova S, Skrha J (2005) Advanced glycoxidation end products in chronic diseases-clinical chemistry and genetic background. Mutat Res 579:37–46
63. Narita T, Sasaki H, Hosoba M, Miura T et al (2004) Parallel increase in urinary excretion rates of immunoglobulin G, ceruloplasmin, transferrin, and orosomucoid in normoalbuminuric type 2 diabetic patients. Diabetes Care 27:1176–1181
64. Horvath AJ, Forsyth SL, Coughlin PB (2004) Expression patterns of murine antichymotrypsin-like genes reflect evolutionary divergence at the Serpina3 locus. J Mol Evol 59:488–497
65. Suzuki Y, Yoshida K, Honda E, Sinohara H (1991) Molecular cloning and sequence analysis of cDNAs coding for guinea pig alpha 1-antiproteinases S and F and contrapsin. J Biol Chem 266:928–932
66. Burkle A (2001) Poly(APD-ribosyl)ation, a DNA damage-driven protein modification and regulator of genomic instability. Cancer Lett 163:1–5
67. Lakin ND, Jackson SP (1999) Regulation of p53 in response to DNA damage. Oncogene 18:7644–7655
68. Kriengsinyos W, Rafii M, Wykes LJ, Ball RO, Pencharz PB (2002) Long-term effects of histidine depletion on whole-body protein metabolism in healthy adults. J Nutr 132:3340–3348
69. Uchida K, Kawakishi S (1993) 2-Oxo-histidine as a novel biological marker for oxidatively modified proteins. FEBS Lett 332:208–210
70. Moulder JE (2002) Report on an interagency workshop on the radiobiology of nuclear terrorism. Radiat Res 158(1):118–124
71. Moulder JE, Medhora M (2011) Advances in mitigation of injuries from radiological terrorism or nuclear accidents. Defence Sci J 61(2):99–104
72. Pellmar TC, Rockwell S (2005) Priority list of research areas for radiological nuclear threat countermeasures. Radiat Res 163(1):115–123
73. Brown SL, Kolozsvary A, Liu J, Jenrow KA, Ryu S, Kim JH (2010) Antioxidant diet supplementation starting 24 hours after exposure reduces radiation lethality. Radiat Res 173(4):462–468
74. MacVittie TJ, Farese AM, Jackson W (2005) Defining the full therapeutic potential of recombinant growth factors in the post radiation-accident environment: the effect of supportive care plus administration of G-CSF. Health Phys 89(5):546–555
75. Epperly MW, Wang H, Jones JA, Dixon T, Montesinos CA, Greenberger JS (2011)

Antioxidant-chemoprevention diet ameliorates late effects of total-body irradiation and supplements radioprotection by MnSOD-plasmid liposome administration. Radiat Res 175(6):759–765

76. Ghosh SN, Zhang R, Fish BL, Semenenko VA, Li XA, Moulder JE et al (2009) Renin-angiotensin system suppression mitigates experimental radiation pneumonitis. Int J Radiat Oncol Biol Phys 75(5):1528–1536

77. Jenrow KA, Brown SL, Liu J, Kolozsvary A, Kim JH (2010) Ramipril mitigates radiation-induced impairment of neurogenesis in the rat dentate gyrus. Radiat Oncol 5(1):Article 6

78. Cohen EP, Bedi M, Irving AA, Jacobs ER, Tomic R, Klein JP et al (2012) Mitigation of late renal and pulmonary injury after hematopoietic stem cell transplantation. Int J Radiat Oncol Biol Phys 83(1):292–296

79. Moulder JE, Robbins MEC, Cohen EP, Hopewell JW, Ward WF (1998) Pharmacologic modification of radiation-induced late normal tissue injury. Cancer Treat Res 93:129–151

80. Bertho JM, Roy L, Souidi M, Benderitter M, Gueguen Y, Lataillade JJ et al (2008) New biological indicators to evaluate and monitor radiation-induced damage: an accident case report. Radiat Res 169(5):543–550

Effects of Radiofrequency-Modulated Electromagnetic Fields on Proteome

6

Dariusz Leszczynski

Abstract

Proteomics, the science that examines the repertoire of proteins present in an organism using both high-throughput and low-throughput techniques, might give a better understanding of the functional processes ongoing in cells than genomics or transcriptomics, because proteins are the molecules that directly regulate physiological processes. Not all changes in gene expression are necessarily reflected in the proteome. Therefore, using proteomics approaches to study the effects of RF-EMF might provide information about potential biological and health effects. Especially that the RF-EMF used in wireless communication devices has very low energy and is unable to directly induce gene mutations.

Keywords

Proteome • Protein expression • Protein activity • Non-ionizing radiation • RF-EMF • Radiofrequency-modulated • Electromagnetic fields • Two-dimensional gel electrophoresis • 2DE • 2DE-DIGE • Phosphorylation • Signaling pathways • Stress response • Stress proteins • Heat-shock proteins

Proteomics approaches, especially these using HTSTs, seem to be particularly suited for elucidation of the effects of RF-EMF because they could reveal effects that are not possible to predict, based on the present, still very limited, knowledge about the biological effects of RF-EMF. The proposed usefulness of proteomics and transcriptomics in search for molecular targets of RF-EMF [1] has been subsequently demonstrated in a pilot 5-step feasibility study [2].

In spite of the potential to discover molecular targets of RF-EM, the progress in research using proteomics is slow, hampered predominantly by the lack of funding and a very limited number of research groups involved.

6.1 Human Volunteer Study

To date, only one study examined effects of RF-EMF on proteome in humans [3]. As the authors stated, this was only a pilot study aimed

D. Leszczynski (✉)
STUK – Radiation and Nuclear Safety Authority,
Laippatie 4, Helsinki 00880, Finland
e-mail: dariusz.leszczynski@stuk.fi

at demonstrating the feasibility of the proteomics approach to study effects of RF-EMF in human skin. Because of the variability of proteomes between individual volunteers the authors have used samples of skin from the same person as the unexposed sham controls. This way they were able to perform pilot study using only ten volunteers. However, use of the 2-DE and silver staining method (only method available at that time in authors' laboratory) and the low number of volunteers in this pilot study diminish the reliability of the changes found in the proteome. In total there were found changes in expression of eight proteins, out of which changes in two of them were observed in all exposed volunteers. Unfortunately, due to technical difficulties, the authors were unable to identify the affected proteins. This study should be repeated with larger number of volunteers and using 2DE-DIGE method to obtain more reliable information whether mobile phone radiation affects proteome of exposed tissues in human body. Interestingly, in spite of the continuing discussion whether RF-EMF radiation can cause any health effects to people, it is still not known whether human body responds to mobile phone radiation on molecular level.

6.2 Animal Model Studies

6.2.1 *Drosophila melanogaster*

Three studies were executed using *Drosophila melanogaster* biological model. In the first one, Weisbrot and co-workers have determined that RF-EMF radiation increases expression of hsp70.1, phosphorylation of ELK-1 kinase and binding activity of the serum response element (SRE) [4]. However, the reliability of the results of this study is weakened because the exposure was performed using antenna of a regular mobile phone, what makes exposure dosimetry unreliable. Lee and co-workers [5] have shown that exposures activate stress response kinases ERK (at 1.6 W/kg) and JNK (at 4.0 W/kg) but not the p38 MAPK. What strengthens the results of this study is that the protein activity results are supported by the gene expression results determined in the same study. Finally, Chavdoula and co-workers [6] have examined the organization of the actin network and found that exposures cause increase in disorganization of the network. However, also in this study exposure dosimetry is unreliable because the authors have used the actual mobile phone. There are no studies using HTST performed in *Drosophila melanogaster* model.

6.2.2 Mouse Models

Several studies were performed in mouse model, using mice of different age (fetus or adult 6–8 weeks of age) and of different strains (C57BL/6N, C57BL/6NTac, hsp70.1 deficient, Balb/c, ICR). In two studies mice strain and age was not specified [7, 8]. Detection of protein expression changes was done by immunocytochemistry using both monoclonal and polyclonal antibodies. Six of the published studies came from the research group of John Finnie in Australia [7–12]. Most of them are based on the same biological material that was separately stained in order to detect different proteins. All-in-all studies from Finnie and co-workers have shown that mobile phone radiation has no effect on the expression of the following proteins: c-fos in adult and in fetal mice brain, stress proteins in fetal brain (Hsp25, Hsp32, Hsp70) and on aquaporin-4 in adult brain and on ionized calcium binding adaptor molecule Iba1. These studies, however, have some problem of reliability because in number of them is missing numerical and statistical analysis and the conclusions are based on only a brief verbal description of immunocytochemistry staining. Lee and co-workers [13] have determined that mobile phone radiation has no effect on the expression of stress proteins (HSP90, HSP70, HSP25) or phosphorylation of stress kinases (ERK, JNK, p38MAPK) in a hsp70.1 deficient mice. The same research group has also examined expression of PCNA, GFAP and NeuN proteins in C57BL/6N mice and found no effects of radiation exposure. However, the results are difficult to appreciate because only visual evaluation was performed in the second study and without any statistics [14]. Maskey and co-workers [15] have shown that mobile phone radiation might affect expression of calbindin and calretinin in different areas of brain.

Only a single study in mouse model used HTST to examine effects of RF-EMF on proteome. Research team from Greece [16] published a study suggesting that RF-EMF alters expression of over 100 proteins in mice brain. The authors speculated that some of the affected proteins are important in regulation of learning, memory and in regulation of processes leading to Alzheimer's disease and suggested far reaching health hazard implications of their observation. However, the only effect that was somewhat shown by Fragopoulou and co-workers is the possibility that RF-EMF might alter expression levels of some brain proteins [16]. The indication of the processes that might be affected was only an unconfirmed speculation because the data concerning the affected proteins, obtained from the proteomic analysis, were insufficiently confirmed. Certain proteins were named as affected based only on identification of protein spots in 2D-gel analyses but the confirmation experiments, using e.g. western blot, that the change is real were not performed.

6.2.3 Rat Models

Several studies were performed using rats of different age (newborn to adult) and of different strains (Wistar, Fisher 344, hairless rat, Sprague–Dawley) and different tissues were examined (brain, skin, kidney, testis, thyroid). Detection of protein expression changes was done mostly by immunocytochemistry using both monoclonal and polyclonal antibodies and in some studies by western blot.

Four studies looked at the effects in rat brain. In three of them was observed no effect on protein expression [17–19] whereas in two of them was observed an effect [19, 20]. Unfortunately, in some of the studies samples were analyzed only visually and without calculating statistical significance and therefore diminishing reliability of the obtained results. In three studies effects of mobile phone radiation on skin of hairless mice were analyzed [21–23]. No effects were observed on any of the analyzed proteins, however, the reliability of some analyzes is questionable. Study by Pyrpasopoulou and co-workers [24] examined effects of mobile phone radiation on kidneys of newborn rats and found by using two methods (immunocytochemistry and in situ hybridization) that exposure affected expression of bone morphogenic protein (BMP-4) and bone morphogenic protein receptors (BMPR-II, BMPR-IA). These observations are strengthened by the similar changes observed by the authors in the expression of the corresponding genes. Esmekaya and co-workers [25] have observed changes in expression of apoptosis-regulating proteins cacpase-3 and caspase-9. However, the results are unreliable because the expression changes were evaluated only by visual examination. Lee and co-workers [26] have examined effects of mobile phone radiation on rat testis and found lack of statistically significant effect on several tested apoptosis associated proteins (p21, p53, bcl-2, caspase-3, PARP).To date, no rat model proteome studies were performed with the use of HTST.

6.3 In Vitro Studies in Human Primary Cells and in Cell Lines

Small number of in vitro studies used HTST to examine effects of RF-EMF. However, because of the variety of limiting factors, caused by the differences in the study designs and/or methods used, the conclusions of all studies should be looked at with caution. As of now it appears that all the studies can be considered rather as "feasibility studies", paving the way for further, more thorough and better designed, proteomics studies.

In all studies was used variety of RF-EMF signals and exposure conditions. Some of the exposure conditions caused changes in protein expression and some not. However, because of very limited number of studies, it is not possible to determine whether any of the observed effects can be in any particular way correlated, or not, with certain exposure conditions.

In all of the studies, proteome changes were determined by separating proteins with two-dimensional electrophoresis (2-DE). The protein spot detection was done in most of the studies by silver staining [27–35]. The silver-based staining procedure for protein visualization, although a well established and widely accepted method, might be not the best option for differential proteomics. This is because of the poor reproducibility of this procedure and the reduced linear dynamic protein concentration range that makes it difficult to draw clear conclusions about changes in protein expression. Nevertheless, the data from 2D-PAGE analyses show that proteome changes caused by RF-EMF do not result in major changes of high abundant proteins, which should be taken into account for designing future proteome studies on RF-EMF effects.

A further limitation of the reliability of some studies is the low number of gel replicates. Normally, the silver-based approach requires high numbers (ten or more) of gel replicates. However, in three of the studies the authors rely on only three gel replicates [30, 31, 35], which is not sufficient to draw any reliable conclusions and may lead to false negative or false positive results. In three studies with sufficient number of replicates were obtained two opposite results suggesting existence of effect of 900 MHz GSM exposure on protein expression [28, 29] but indicating lack of effect when exposing to 1,800 MHz GSM signal [32].

The use of relatively novel DIGE system for 2-DE (2-D Fluorescence Difference Gel Electrophoresis), currently considered to be the gold standard in 2-DE, might have been limited in mobile phone radiation proteomics studies by the availability of specialized hardware in the laboratories performing the studies and the high costs for the reagents. There is only a single study [33] where changes in protein expression were examined using 2-DE-DIGE system and with a sufficiently high number of ten replicate gels. This study has shown that the exposure of human primary endothelial cells at SAR of 2.0 W/kg for 1 h using an 1,800 MHz GSM signal does not cause changes in protein expression. Because of the techniques used, this is the most reliable differential proteome analysis to date.

An interesting hypothesis has been presented in the study by Gerner and co-workers [34]. The authors suggest that the protein expression changes might be not the best end-point to show effects of RF-EMF. They have demonstrated that in the cells exposed to mobile phone radiation changes de novo synthesis of some proteins, without statistically significant change in the overall protein expression. However, these conclusions are weakened by only three gel replicates performed in the study, what is not sufficient for reliable determination of changes in protein expression.

The observations of Gerner and co-workers [34] and Leszczynski and co-workers [27] suggest that examining of only protein expression is not sufficient to detect effects of mobile phone radiation. Experiments that examine de novo protein synthesis as well as post-translational modifications of proteins (e.g. phosphorylation) are necessary in order to determine impact low-energy RF-EMF exposures on cell proteome.

The numbers of proteins shown to be affected by RF-EMF, in all proteomics studies that show an effect, are not only low but they are lower than the number of expected false positives. This argument is often used to suggest that there is no effect of RF-EMF on cell proteome. However, the low number of responding proteins, even when it is below the expected rate of false positives, does not mean automatically that the every protein appearing as affected by RF-EMF is a sure false positive finding. The calculation of the number of expected false positives shows only the probability that the affected proteins might be false positives, but it does not prove that all of them are indeed false positives. The proof can be only obtained by biological experiment. This has been demonstrated in study where the number of detected statistically significantly affected proteins was lower than the expected number of false positives [28]. However, further western blot analysis of the changes in the expression of one of the affected proteins, vimentin, has shown that this protein did respond to RF-EMF.

6.4 Conclusions

It is necessary to keep in mind that the execution of a screening study using proteomics, or other high-throughput screening approach, is just the beginning of the process of finding out the biological effects of RF-EMF. The validity of expression changes detected in identified protein targets needs to be confirmed by other, non-high-throughput screening methods. Such validation experiments were performed only in four of the nine published proteomics studies. In two of them [27, 28] the authors were able to confirm that the proteomics identified targets were correct. Whereas in two other studies, were not [32, 35]. However, the success of the validation experiments depends e.g. on the availability of antibodies directed against the identified protein targets [32]. Once the affected proteins are confirmed, this information can be used for determining what physiological functions of the cell might be altered by the observed protein expression change and whether the observed change is of sufficient magnitude to change cell physiology. Only when the altered protein expression is able to alter cell physiology we can suspect that the effect might have some potential to cause biological or health-related effects but if the cell physiology will not be affected then there will not be any risk of health effects.

In summary, because of the small number of executed proteomics studies, with the variety of shortcomings caused by study design and by the availability of methods and reagents, the information provided by them is very limited, at the best. In the future, proteomics studies should be continued [36] in order to get sufficient amount of information that will permit to draw valid conclusions about the impact of RF-EMF on cell proteome and, consequently, on cell physiology.

References

1. Leszczynski D, Joenväärä S (2001) Proteomics: new way to determine possible biological effects of mobile phone radiation. Nat Genet 27(Suppl):67
2. Leszczynski D, Nylund R, Joenväärä S, Reivinen J (2004) Applicability of discovery science approach to determine biological effects of mobile phone radiation. Proteomics 4:426–431
3. Karinen A, Heinävaara S, Nylund R, Leszczynski D (2008) Mobile phone radiation might alter protein expression in human skin. BMC Genomics 9:77–81
4. Weisbrot D, Lin H, Ye L, Blank M, Goodman R (2003) Effects of mobile phone radiation on reproduction and development in *Drosophila melanogaster*. J Cell Biochem 89:48–55
5. Lee KS, Choi JS, Hong SY, Son TH, Yu K (2008) Mobile phone electromagnetic radiation activates MAPK signaling and regulates viability in Drosophila. Bioelectromagnetics 29:371–379
6. Chavdoula ED, Panagopoulos DJ, Margaritis LH (2010) Comparison of biological effects between continuous and intermittent exposure to GSM-900-MHz mobile phone radiation: detection of apoptotic cell-death features. Mutat Res 700:51–61
7. Finnie JW, Blumbergs PC, Cai Z, Manavis J (2009) Expression of the water channel protein, aquaporin-4, in mouse brains exposed to mobile phone radiofrequency fields. Pathology 41:473–475
8. Finnie JW, Cai Z, Manavis J, Helps S, Blumbergs PC (2010) Microglial activation as a measure of stress in mouse brains exposed acutely (60 minutes) and long-term (2 years) to mobile telephone radiofrequency fields. Pathology 42:151–154
9. Finnie JW (2005) Expression of the immediate early gene, c-fos, in mouse brain after acute global system for mobile communication microwave exposure. Pathology 37:231–233
10. Finnie JW, Cai Z, Blumbergs PC, Manavis J, Kuchel TR (2006) Expression of the immediate early gene, c-fos, in fetal brain after whole gestation exposure of pregnant mice to global system for mobile communication microwaves. Pathology 38:333–335
11. Finnie JW, Cai Z, Blumbergs PC, Manavis J, Kuchel TR (2007) Stress response in mouse brain after long-term (2 year) exposure to mobile telephone radiofrequency fields using the immediate early gene, c-fos. Pathology 39:271–273
12. Finnie JW, Chidlow G, Blumbergs PC, Manavis J, Cai Z (2009) Heat shock protein induction in fetal mouse brain as a measure of stress after whole of gestation exposure to mobile telephony radiofrequency fields. Pathology 41:276–279
13. Lee JS, Huang TQ, Lee JJ, Pack JK, Jang JJ, Seo JS (2005) Subchronic exposure of hsp70.1-deficient mice to radiofrequency radiation. Int J Radiat Biol 81:781–792

14. Kim TH, Huang TQ, Jang JJ, Kim MH, Kim HJ, Lee JS, Pack JK, Seo JS, Park WY (2008) Local exposure of 849 MHz and 1763 MHz radiofrequency radiation to mouse heads does not induce cell death or cell proliferation in brain. Exp Mol Med 40: 294–303
15. Maskey D, Kim M, Aryal B, Pradhan J, Choi IY, Park KS, Son T, Hong SY, Kim SB, Kim HG, Kim MJ (2010) Effect of 835 MHz radiofrequency radiation exposure on calcium binding proteins in the hippocampus of the mouse brain. Brain Res 1313: 232–241
16. Fragopoulou AF, Samara A, Antonelou MH, Xanthopoulou A, Papadopoulou A, Vougas K, Koutsogiannopoulou E, Anastasiadou E, Stravopodis DJ, Tsangaris GT, Margaritis LH (2012) Brain proteome response following whole body exposure of mice to mobile phone or wireless DECT base radiation. Electromagn Biol Med. doi:10.3109/15368378.2011.631068
17. Fritze K, Wiessner C, Kuster N, Sommer C, Gass P, Hermann DM, Kiessling M, Hossmann DK (1997) Effect of global system for mobile communication microwave exposure on the genomic response of the rat brain. Neuroscience 81:627–639
18. Belyaev IY, Baureus Koch C, Terenius O, Roxström-Lindquist K, Malmgren LOG, Sommer WH, Salford LG, Persson BRR (2006) Exposure of rat brain to 915 MHz GSM microwaves induces changes in gene expression but not double stranded DNA breaks or effects on chromatin conformation. Bioelectromagnetics 27:295–306
19. Dasdag S, Akdag MZ, Ulukaya E, Uzunlar AK, Ocak AR (2009) Effect of mobile phone exposure on apoptotic glial cells and status of oxidative stress in rat brain. Electromagn Biol Med 28:342–354
20. Ammari M, Gamez C, Lecomte A, Sakly M, Abdelmelek H, De Seze R (2010) GFAP expression in the rat brain following sub-chronic exposure to a 900 MHz electromagnetic field signal. Int J Radiat Biol 86:367–375
21. Masuda H, Sanchez S, Dulou PE, Haro E, Anane R, Billaudel B, Leveque P, Veyret B (2006) Effect of GSM-900 and -1800 signals on the skin of hairless rats. I: 2-hour acute exposures. Int J Radiat Biol 82:669–674
22. Sanchez S, Masuda H, Billaudel B, Haro E, Anane R, Leveque P, Ruffie G, Lagroye I, Veyret B (2006) Effect of GSM-900 and -1800 signals on the skin of hairless rats. II: 12-week chronic exposures. Int J Radiat Biol 82:675–680
23. Sanchez S, Masuda H, Ruffie G, Poulletier De Gannes F, Billaudel B, Haro E, Leveque P, Lagroye I, Veyret B (2008) Effect of GSM-900 and -1800 signals on the skin of hairless rats. III: Expression of heat shock proteins. Int J Radiat Biol 84:61–68
24. Pyrpasopoulou A, Kotoula V, Cheva A, Hytiroglou P, Nikolakaki E, Magras IN, Xenos TD, Tsiboukis TD, Karkavelas G (2004) Bone morphogenetic protein expression in newborn rat kidneys after prenatal exposure to radiofrequency radiation. Bioelectromagnetics 25:216–227
25. Esmekaya MA, Seyhan N, Ömeroglu S (2010) Pulse modulated 900 MHz radiation induces hypothyroidism and apoptosis in thyroid cells: a light, electron microscopy and immunohistochemical study. Int J Radiat Biol 86:1106–1116
26. Lee HJ, Pack JK, Kim TH, Kim N, Choi SY, Lee JS, Kim SH, Lee YS (2010) The lack of histological changes of CDMA cellular phone-based radio frequency on rat testis. Bioelectromagnetics 31:528–534
27. Leszczynski D, Joenväärä S, Reivinen J, Kuokka R (2002) Non-thermal activation of hsp27/p38MAPK stress pathway by mobile phone radiation in human endothelial cells: molecular mechanism for cancer- and blood–brain barrier-related effects. Differentiation 70:120–129
28. Nylund R, Leszczynski D (2004) Proteomics analysis of human endothelial cell line EA.hy926 after exposure to GSM 900 radiation. Proteomics 4:1359–1365
29. Nylund R, Leszczynski D (2006) Mobile phone radiation causes changes in gene and protein expression in human endothelial cell lines and the response seems to be genome- and proteome-dependent. Proteomics 6:4769–4780
30. Zeng Q, Chen G, Weng Y, Wang L, Chiang H, Lu D, Xu Z (2006) Effects of global system for mobile communications 1800 MHz radiofrequency electromagnetic fields on gene and protein expression in MCF-7 cells. Proteomics 6:4732–4738
31. Li HW, Yao K, Jin HY, Sun LX, Lu DQ, Yu YB (2007) Proteomic analysis of human lens epithelial cells exposed to microwaves. Jpn J Ophtalmol 51:412–416
32. Nylund R, Tammio H, Kuster N, Leszczynski D (2009) Proteomic analysis of the response of human endothelial cell line EA.hy926 to 1800 GSM mobile phone radiation. J Proteomic Bioinform 2:455–462
33. Nylund R, Kuster N, Leszczynski D (2010) Analysis of proteome response to the mobile phone radiation in two types of human primary endothelial cells. Proteome Sci 8:52–58
34. Gerner C, Haudek V, Schnadl U, Bayer E, Gundacker N, Hutter HP, Mosgoeller W (2010) Increased protein synthesis by cells exposed to a 1800 MHz radiofrequency mobile phone electromagnetic field detected by proteome profiling. Int Arch Occup Environ Health 83:691–702
35. Kim KB, Byun HO, Han NK, Ko YG, Choi HD, Kim N, Pack JK, Lee JS (2010) Two-dimensional electrophoretic analysis of radio frequency radiation-exposed MCF7 breast cancer cells. J Radiat Res 51:205–213
36. Leszczynski D, Meltz ML (2006) Report: questions and answers concerning applicability of proteomics and transcriptomics in EMF research. Proteomics 6:4674–4677

Global Protein Expression in Response to Extremely Low Frequency Magnetic Fields

Guangdi Chen and Zhengping Xu

Abstract

Daily exposure to extremely low frequency magnetic fields (ELF MF) in the environment has raised public concerns on human health. Epidemiological studies suggest that exposure to ELF MF might associate with an elevated risk of cancer and other diseases in humans. To explain and/or support epidemiological observations, many laboratory studies have been conducted to elucidate the biological effects of ELF MF exposure and the underlying mechanisms of action. In order to reveal the global effects of ELF MF on protein expression, the proteomics approaches has been employed in this research field. In 2005, WHO organized a Workshop on Application of Proteomics and Transcriptomics in electromagnetic fields (EMF) Research in Helsinki, Finland to discuss the related problems and solutions. Later the journal Proteomics published a special issue devoted to the application of proteomics to EMF research. This chapter aims to summarize the current research progress and discuss the applicability of proteomics approaches in studying on ELF MF induced biological effects and the underlying mechanisms.

Keywords

Proteome • Protein expression • Non-ionizing radiation • Electromagnetic fields • EMF • Extremely low frequency magnetic fields • ELF-MF • Two-dimensional gel electrophoresis • Mass spectrometry • Yeast

7.1 Introduction

Daily exposure to electromagnetic fields (EMF), including extremely low frequency magnetic fields (ELF MF) in the environment has raised

G. Chen, Ph.D. (✉) • Z. Xu, Ph.D.
Bioelectromagnetics Laboratory, Zhejiang University School of Medicine, Hangzhou 310058, China
e-mail: chenguangdi@gmail.com; zpxu@zju.edu.cn

public concerns about whether they have harmful consequences on human health. Several epidemiological studies suggest that exposure to EMF might associate with an elevated risk of cancer and other diseases in humans (reviewed in Feychting et al. [1]). To explain and/or support epidemiological observations, many laboratory studies have been conducted, but the results were controversial and no clear conclusion could be drawn to assess EMF health risk.

It is reasoned that one of the priorities in EMF research is to elucidate the biological effects of EMF exposure and the underlying mechanisms of action. Proteins are key players in organisms, and it has been assumed that any biological impact of EMF must be mediated by alterations in protein expression [2, 3]. For example, heat shock proteins have been identified as EMF responsive genes and/or proteins in certain biological systems [4]. In order to reveal the global effects of EMF on protein expression, transcriptomics and proteomics, as high-throughput screening techniques (HTSTs), were eventually employed in EMF research with an intention to screen potential EMF-responsive genes and/or proteins without any bias. In 2005, WHO organized a Workshop on Application of Proteomics and Transcriptomics in EMF Research in Helsinki, Finland to discuss the related problems and solutions in this field [5, 6]. Later the journal Proteomics published a special issue devoted to the application of proteomics and transcriptomics to EMF research. This review aims to summarize the current research progress and discuss the applicability of proteomics approaches in the investigations on ELF MF induced biological effects and the underlying mechanisms.

7.2 Model Organism

Nakasono et al. has investigated the effects of protein expression in model system such as *Escherichia coli* and *Saccharomyces cerevisiae* using two dimensional gels electrophoresis (2-DE) method. When the bacterial cells were exposed to each MF at 5–100 Hz under aerobic conditions (6.5 h) or at 50 Hz under anaerobic conditions (16 h) at the maximum intensity (7.8–14 mT), no reproducible changes were observed in the 2D gels. However, the stress-sensitive proteins did respond to most stress factors, including temperature change, chemical compounds, heavy metals, and nutrients. The authors concluded that the high-intensity ELF MF (14 mT at power frequency) did not act as a general stress factor [7]. When using *Saccharomyces cerevisiae* as a model system, Nakasono et al. reported that no reproducible changes in the 2D gels were observed in yeast cells after exposure to 50 Hz MF at the intensity up to 300 mT for 24 h [8]. In this study, only three sets of gels from three independent experiments were analyzed.

Using 2-D Fluorescence Difference Gel Electrophoresis (2-D DIGE) technology and mass spectrum (MS) in a blind study, Sinclair et al. have investigated the effects of ELF MF on the proteomes of wild type *Schizosaccharomyces pombe* and a Sty1p deletion mutant which displays increased sensitivity to a variety of cellular stresses [9]. The yeast cells were exposed to 50 Hz EMF at field strength of 1 mT for 60 min. While this study identified a number of protein isoforms that displayed significant differential expressions across experimental conditions, there was no correlation between their patterns of expression and the ELF MF exposure regimen. The authors concluded that there were no significant effects of ELF MF on the yeast proteome at the sensitivity afforded by 2D-DIGE. They hypothesized that the proteins identified in the experiments must be sensitive to subtle changes in culture and/or handling conditions [9].

7.3 Mammalian Cells

Li et al. have performed a proteomics approach to investigate the changes of protein expression profile induced by ELF MF in human breast cancer cell line MCF-7. With help of 2-DE and data analysis on nine gels for each group, 44 differentially expressed protein spots were screened in MCF-7 cells after exposure to 0.4 mT 50 Hz MF for 24 h. Three proteins were identified by LC-IT Tandem mass spectrum (MS) as RNA binding protein regulatory subunit, proteasome subunit beta type 7 precursor, and translationally controlled tumor protein, respectively [10]. Further investigations, such as Western blotting, are required to confirm these ELF responsive candidate proteins.

Kanitz et al. used proteomic methods to investigate the biochemical effects induced by MF exposure in SF767 human glioma cells [11]. The cells were exposed to 1.2 µT of 60 Hz MF or epidermal growth factor (EGF). SF767 cells were exposed for 3 h to sham conditions (<0.2 µT ambient field strength) or 1.2 µT of MF with or without combination of 10 ng/ml of EGF. The solubilized proteins from four groups of cells (sham; 1.2 µT MF; sham + EGF; 1.2 µT MF + EGF) were loaded for electrophoresis by 2D-PAGE and stained using a colloidal Coomassie blue technique to resolve and characterize the proteins. The spots with significant alterations in the densities were excised and subjected to peptide mass fingerprinting. After exposure to 1.2 µT of MF for 3 h, the mean abundances of ten identified proteins were significantly altered, including three proteins with increased expression level and seven decreased. In the presence of EGF, the MF exposure changed protein expression profile in SF767 cells, and four proteins were identified with increased expression level and two decreased. The authors suggested that differentially expressed proteins in SF767 cells may be useful as biomarkers for biological changes caused by exposure to magnetic fields. However, these candidate MF responsive proteins should be validated and the biological functions need further elucidated.

In the recent study by Sulpizio et al., human SH-SY5Y neuroblastoma cells were exposed to a 50 Hz, 1 mT sinusoidal MF at three different times i.e. 5 days (T5), 10 days (T10) and 15 days (T15) and then the effects of MF exposure on proteome expression were investigated by 2D-PAGE and MS analyses [12]. Through comparative analysis between treated and control samples, the authors analyzed the proteome changes induced by the MF exposure. Nine new proteins resolved in sample after a 15 day treatment, were involved in a cellular defense mechanism and/or in cellular organization and proliferation such as peroxiredoxin isoenzymes (2, 3 and 6), 3-mercaptopyruvate sulfurtransferase, actin cytoplasmatic 2, t-complex protein subunit beta, ropporin-1A and profilin-2 and spindlin-1. The authors also showed that the MF exposure altered the cell proliferation and cell viability. Furthermore, the MF-exposed cells showed a higher and more widespread expression level of alpha tubulin, especially in the periphery of cell clusters, compared to control cells, suggesting that the MF exposure induce a spatial orientation of cells. The authors hypothesized that the MF exposure could trigger a shift toward a more invasive phenotype. However, future studies are needed to address this hypothesis.

7.4 Summary

Generally, recent studies on global protein expression responding to ELF MF have been conducted in different biological systems by applications of different proteomics approaches. The mammalian cells seem more sensitive to ELF MF exposure, however, the bacterial and yeast cells, as model organism, did not react to ELF MF exposure. Some proteome analyses showed that ELF MF exposure could change protein expression in the mammalian cells; there are lacks of confirmations by other assays to determine if they are real ELF MF responsive proteins and future studies are needed to elucidate the biological functions of these candidate ELF MF responsive proteins. The human neuroblastoma cell line SH-SY5Y is suggested as a model to further confirm the effect of ELF MF on global protein expression, and the role of ELF MF responsive proteins in ELF MF induced cell behavior changes.

References

1. Feychting M, Ahlbom A, Kheifets L (2005) EMF and health. Annu Rev Public Health 26:165–189
2. Phillips JL, Haggren W, Thomas WJ, Ishida-Jones T, Adey WR (1992) Magnetic field-induced changes in specific gene transcription. Biochim Biophys Acta 1132(2):140–144
3. Wei LX, Goodman R, Henderson A (1990) Changes in levels of c-myc and histone H2B following exposure of cells to low-frequency sinusoidal electromagnetic fields: evidence for a window effect. Bioelectromagnetics 11(4):269–272
4. Leszczynski D, Joenvaara S, Reivinen J, Kuokka R (2002) Non-thermal activation of the hsp27/p38MAPK stress pathway by mobile phone radiation in human endothelial cells: molecular

mechanism for cancer- and blood–brain barrier-related effects. Differentiation 70(2–3):120–129
5. Leszczynski D (2006) The need for a new approach in studies of the biological effects of electromagnetic fields. Proteomics 6(17):4671–4673
6. Leszczynski D, Meltz ML (2006) Questions and answers concerning applicability of proteomics and transcriptomics in EMF research. Proteomics 6(17):4674–4677
7. Nakasono S, Saiki H (2000) Effect of ELF magnetic fields on protein synthesis in *Escherichia coli* K12. Radiat Res 154(2):208–216
8. Nakasono S, Laramee C, Saiki H, McLeod KJ (2003) Effect of power-frequency magnetic fields on genome-scale gene expression in *Saccharomyces cerevisiae*. Radiat Res 160(1):25–37
9. Sinclair J, Weeks M, Butt A, Worthington JL, Akpan A, Jones N, Waterfield M, Alland D, Timms JF (2006) Proteomic response of *Schizosaccharomyces pombe* to static and oscillating extremely low-frequency electromagnetic fields. Proteomics 6(17):4755–4764
10. Li H, Zeng Q, Weng Y, Lu D, Jiang H, Xu Z (2005) Effects of ELF magnetic fields on protein expression profile of human breast cancer cell MCF7. Sci China C Life Sci 48(5):506–514
11. Kanitz MH, Witzmann FA, Lotz WG, Conover D, Savage RE (2007) Investigation of protein expression in magnetic field-treated human glioma cells. Bioelectromagnetics 28(7):546–552
12. Sulpizio M, Falone S, Amicarelli F, Marchisio M, Di Giuseppe F, Eleuterio E, Di Ilio C, Angelucci S (2011) Molecular basis underlying the biological effects elicited by extremely low frequency magnetic field (ELF-MF) on neuroblastoma cells. J Cell Biochem 112(12):3797–3806

Ultraviolet Radiation Effects on the Proteome of Skin Cells

H. Konrad Muller and Gregory M. Woods

Abstract

Proteomic studies to date have had limited use as an investigative tool in the skin's response to UV radiation. These studies used cell lines and reconstructed skin and have shown evidence of cell injury with oxidative damage and stress induced heat shock proteins. Others changes included altered cytokeratin and cytoskeletal proteins with enhanced expression of TRIM29 as the keratinocytes regenerate. The associated DNA repair requires polη, Rad18/Rad16 and Rev1. In the whole animal these events would be associated with inflammation, remodelling of the epidermis and modulation of the immune response. Longer term changes include ageing and skin cancers such as melanoma, squamous cell carcinoma and basal cell carcinoma. In the future proteomics will be used to explore these important aspects of photobiology. Better characterisation of the proteins involved should lead to a greater understanding of the skin's response to UV radiation.

Keywords

Proteome • Skin • Non-ionizing radiation • Ultraviolet radiation • UV • UVA • UVB • UVC • Acute effects • Sunburn cells • Keratinocyte • Apoptosis • Delayed effects • Tanning • Skin cancer • Melanoma • Immune system • Langerhans cells • Immune suppression

H.K. Muller (✉)
School of Medicine, University of Tasmania, Private Bag 34, Hobart, TAS 7000, Australia
e-mail: Konrad.Muller@utas.edu.au

G.M. Woods
School of Medicine, University of Tasmania, Private Bag 34, Hobart, TAS 7000, Australia

Menzies Research Institute Tasmania, Private Bag 23, Hobart, TAS 7000, Australia
e-mail: G.M.Woods@utas.edu.au

8.1 Background

8.1.1 Introduction

Ultraviolet (UV) radiation produces both beneficial and adverse effects on the skin. The former includes Vitamin D synthesis while adverse effects range from acute to chronic changes including DNA damage, ageing and skin cancer. The skin immune response is also modulated by UV radiation. The morphological events associated with UV radiation have been well characterised and range from sunburn cells in acute injury through to malignancy.

Ultraviolet radiation is part of the electromagnetic spectrum that includes x-rays, cosmic rays and visible light of the sun's emission spectrum. The UV spectrum is divided into three components: UVA (320–400 nm), UVB (280–320 nm) and UVC (200–280 nm). UVC does not normally reach the earth's surface and is filtered out by the ozone layer. UVB has a wide range of both acute and chronic adverse effects and is the major wavelength responsible for skin cancer. UVA is less damaging, being associated with ageing skin events, particularly dermal protein and elastic changes. More recent studies have implicated UVA in cutaneous carcinogenesis, particularly by oxidative DNA damage. Both UVB and UVA have effects on the skin immune system, although UVB has been analysed in more detail and has a greater range of effects than UVA [1].

8.1.2 Acute Effects of Ultraviolet Radiation

The most obvious effect of UV radiation on the skin is acute inflammation associated with sunburn. The initial response in the first 30 min is associated with histamine release. The peak reaction at 24 h, is accompanied by subepidermal blistering and infiltration of the tissue with neutrophilic granulocytes and is mediated by prostaglandins (PGE2) and leukotrienes [2]. At this time proinflammatory cytokines such as IL-1 and IL-6 are released from the damaged tissue as are anti-inflammatory molecules such as IL-4 and IL-10. IL-10 had a profound immunoregulatory role involving regulatory T cells (Tregs) [3].

A morphological feature of sunburn is the presence of sunburn cells in the epidermis. These are apoptotic keratinocytes as a result of DNA damage, which are present by 6 h and peaks at approximately 24 h [4]. The damaged DNA is not repairable and caspases are activated leading to DNA fragmentation and apoptosis. The apoptotic cells are removed by phagocytic cells [5]. Interestingly it appears to be the IL-10-producing macrophages that preferentially remove the apoptotic cells [6].

Endonucleases can repair DNA damage leading to cell survival and allowing epidermal regeneration. Regeneration tends to peak at 48–72 h and the epidermis is now hyperplastic and much thicker than normal. The outer keratin layer is also increased and remodelling of the epidermis follows [4].

Epidermal repair post UV radiation is not an accurate process and leads to DNA mutations. Classically, the UVB radiation causes cross linking of adjacent thymine groups resulting in thymine dimer formation. The dimers are removed by endonucleases and the thymine replaced. However, with mutations a wrong base is inserted converting an AT to a CG pairing. These mutations are retained providing a basis for subsequent neoplastic development.

Another feature of the acute response is sun-tanning or pigmentation due to the production of melanin in the melanocytes. Melanin, a protein produced from tyrosine within melanocytes, is packaged in melanosomes. Sun-tanning may be immediate or delayed. Immediate pigmentation only lasts minutes to hours and is due to a redistribution of the existing melanin into the adjacent keratinocytes. With delayed tanning new melanin is produced, which occurs within hours and the production can last for days. UVB is more effective than UVA in generating new melanin [7]. Delayed tanning may occur without burning after UVA, but not usually after UVB [8].

8.1.3 Chronic Effects of Ultraviolet Radiation

The long term effects of ultraviolet radiation exposure are enhanced skin ageing and skin cancer. Long exposure across life to the sun's ultraviolet radiation increases the ageing process. At the gross level it is seen in the most sun exposed areas such as the face and arms where the skin is wrinkled and there is a loss of texture. On skin biopsies, the dermal connective tissue shows marked degeneration of the collagen and elastic replication, a lesion known as solar elastosis.

The most important long term effects of ultraviolet light are skin cancers. Here we have a spectrum of changes ranging from the premalignant lesion, solar keratosis, through to the malignant lesions, squamous and basal cell carcinomas and malignant melanoma. With solar keratosis there is a hyperplastic lesion with increased keratin production and epidermal downward projection into the dermis. Within these there are areas of cell atypia and loss of cell polarity. It is these particular cellular features that may lead to invasive squamous cell carcinoma. Yet the number of lesions that progress down this path are relatively small. A population study by Marks and colleagues in 1988 reported that the risk of solar keratosis progressing to squamous cell carcinoma within 1 year was less than 1 per 1,000 [9].

With both squamous and basal cell carcinoma there is local invasion of the dermis with malignant keratinocytes in the former which can still produce keratin as we see in keratin pearls in the centre of well differentiated tumours. With basal cell carcinoma these arise from cells in the basal epidermal layer. A feature of this tumour is its local destructive capacity if not excised. They can invade bone and muscle if not treated [10].

Melanoma arises from malignant melanocytes also found in the basal epidermal layer. While pathologists recognise various patterns of melanoma growth, they have the capacity to spread widely in the body and metastasise to all tissues including cardiac muscle. A challenge for the pathologist remains the problem of atypical melanocytic lesions arising in the basal epidermis; these remain diagnostic challenges.

Ultraviolet radiation is a causative agent in all these malignant skin tumours. With squamous cell carcinoma there is repeated long term exposure so that these lesions are seen particularly in the sun exposed areas of outdoor workers. With basal cell carcinoma there is evidence of both acute exposure in early life [11] and long term exposure on sunlight exposed areas [12, 13].

Melanoma can arise from either non-exposed or ultraviolet radiation exposed skin. With the former these may be preceded by pigmented naevi [14, 15]. However, sunburn doses of ultraviolet light in childhood are clearly linked to melanoma developing in adult life. In this situation initiated precursors of melanoma remain in the skin for years. Our knowledge of the events at this point is poorly understood. Attempts to analyse this problem have used transgenic mice expressing either HGF [16, 17] or TPras [18, 19]. Such mice exposed to ultraviolet light in the neonatal period at 3 days develop malignant melanoma by 4 months at the earliest. What is also evident at this early time in the neonatal period is immune tolerance [20]. So, in this setting precursor cells for melanoma are being established at a time when there is limited immunosurveillance [21].

8.1.4 Ultraviolet Radiation and the Skin Immune System

Ultraviolet radiation of skin not only results in cutaneous cancers but also induces immunosuppression in human and animal models. The most detailed analysis has been carried out in mice. UVB causes both systemic and local immunosuppression. In the former, when antigen is applied to UVB treated skin at a site distant to that of UV exposure, immunosuppression of the whole animal follows within 24 h. With local immunosuppression antigen is applied directly to the site of UV treatment, again leading to down regulation of the immune response. This local immunosuppression is associated with depletion of Langerhans cells (LC) from the skin so that antigen presented at this site is processed by newly arrived LC [20]. Local and systemic suppression are associated with the generation of Tregs. Their transfer into mice has formally demonstrated this either before or after they have been exposed to specific antigen [22, 23]. Specificity of the immunosuppression is central to this concept.

The induction of these immunosuppressive events, particularly in local immunosuppression, involves altered LC and cytokines produced by keratinocytes and mast cells. These include prostaglandins, cis-urocanic acid (UCA) and IL-10 [2]. The receptor for the UCA is 5 hydroxytryptamine [24].

The release of immunomodulating molecules from mast cells post UVB also involves neuropeptides such as substance P and calcitonin gene related protein (CGRP) released from the axon reflex [25–27]. These trigger degranulation of mast cells and the production of molecules such as IL-10 which limits skin pathology associated with UVB radiation [28] and suppresses germinal centre formation [29].

With local immunosuppression the altered LC have modified antigen presentation so that antigen specific regulatory T cells (Tregs) are induced. In contrast antigen presenting cells are not altered by UVB induced systemic suppression. In this case the immunosuppression is mediated by IL-10 and/or cis-urocanic acid and the generation of specific Tregs is associated with the induction of rank ligand [30]. More recent experiments by Byrne and Halliday have implicated regulatory B cells in this systemic response post UVB exposure [31–33]. In addition, UVB induced DNA changes have also been described as a key step leading to the production of suppressive cytokines [34, 35].

The studies of Kripke and Fisher in the early 1970s provided the first link between UV induction and growth of skin tumours and immunosuppressive events. While these squamous cell tumours were highly antigenic and could be rejected in normal mice, in those that had been treated with UVB the tumours were not rejected. In fact, T suppressor cells as they were called developed in UV radiated mice before the development of tumours [35a]. Failure to mount an effective immune response to tumours allows their growth and metastatic spread [21, 36].

The above is a broad overview highlighting the major acute and chronic effects of UVB radiation on the skin prior to the advent of proteomics. The use of proteomics now offers the opportunity to more precisely characterise key events in the skin after UVB radiation. These can be based on skin specimens obtained from whole animal studies or using in vitro systems and cultured cells.

8.2 UV Radiation and Proteomics

Review of the literature reveals few studies in this area and most have involved cultured cells or artificial skin.

For comparison the normal skin proteome needs to be available so that changes induced by UV radiation can be evaluated. The only studies we are aware of on normal skin are on murine epidermis [37].

Our own study on neonatal murine epidermis identified 179 protein differences between neonatal (day 4) and mature (day 21) skin. Of particular interest was the overexpression of stefin A in neonatal epidermis suggesting that this protein has an important role in development and potentially the neonatal immune response [38]. This approach provided an opportunity to identify a range of proteins that are altered in different situations. One ideal situation would be a comparative study of UV treated and normal skin.

With large doses of UV radiation skin burning occurs with apoptosis of keratinocytes and impairment of protein production [39]. Besides interacting with DNA, UV also interacts with cellular chromophores and photosensitizers with the generation of reactive oxygen species that cause oxidative damage. This triggers cellular signalling pathways related to growth differentiation, senescence, dermal connective tissue degradation and inflammation [40, 41]. Some of these events have been analysed in the following studies, which we now summarise.

Repeated sub lethal doses of UVB induce an alternative differentiation state rather than cell death and senescence of cultivated human keratinocytes lacking functional $p16^{INK-4a}$ immortalised with telomerase and retaining their ability to differentiate [42, 43]. Bertrand-Vallery and colleagues used these cells called N-hTERT to perform a proteomic analysis after repeated doses of UVB; doses of 300 mJ/cm^2 per exposure were used. Proteomic profiling with fluorescent two dimensional differences in-gel electrophoresis (2D-DIGE) demonstrated some 69 proteins following up to 64 h post UVB treatments [44].

In these experiments UVB induced an increased abundance of involucrin, a late marker of differentiation and cytokeratins K6, K16 and K17 [42]. Other proteins found were plasminogen activator inhibitor–2 (PAI-2), 14-3-3-6, phosphorylation of p38^{MAPK} and heat shock protein 27 (HSP27). Other proteins not previously described in abundance after UVB were TRIpartite Motif Protein 29 (TRIM29), Cap G and cytokeratin 8 (K8). They also found an elevated secretion of metalloproteinase-9 as seen in primary keratinocytes in vivo in the epidermis [40, 45, 46].

Cap G is an actin-capping protein involved in cell signalling, receptor mediated membrane ruffling, phagocytosis and cell mobility [40, 45, 46]. Actin binding proteins are among cytoskeletal proteins regulated by UV [47]. S73 phosphorylation of K8 is a stress response [48]. Such post translational modifications could promote the survival of keratinocytes after repeated doses of UVB.

Bertrand-Vallery et al. showed a correlation between TRIM29 abundance and cell survival after exposure to UVB. The ERK pathway is linked to cell survival [49]. Activated p38^{MAPK} contributes to the activation of numerous transcription factors associated with cell cycle arrest, differentiation, apoptosis and inflammation in injured tissues [50].

UVB-induced increase of TRIM 9 mRNA and protein is regulated by PKC8, which can phosphorylate the TRIM29 protein. The expression of TRIM29 can be knocked down by shRNA decreasing survival of N-hTERT cells after UV treatment. This confirms the functional role of TRIM29 [44].

Bertrand-Vallery et al. concluded that TRIM29 allows keratinocytes to enter an alternative survival differentiation process rather than die after UVB induced stress [44]. These studies contrast with higher doses of UVB where apoptosis and sunburn cells are induced via caspases [51]. However, cells that survive UV radiation are likely to carry mutations that activate malignant pathways.

Hensbergen and colleagues used three dimensional human skin cultures, representing a substitute for human skin, to examine the proteomic profile of these cells after exposure to solar simulated light (96% UVA, 4% UVB) [52]. These epidermal constructs retain a functional stratum corneum, which enables UV doses to be used similar to human sunlight exposure. They used two dimensional polyacrylamide gel electrophoresis (2D-PAGE) to detect differential expressions of proteins. Their analysis defined 11 proteins, which were further examined by mass spectrometric analysis. Prominent among these were heat shock protein (HSP27), superoxide dismutase (MnSOD) and PDX-2 and 60S acidic ribosomal phosphoprotein PO, which are known to be involved in UV responses. They identified two novel proteins, cofilin-1 and destrin, which were down regulated. On further analysis these proteins were de-phosphorylated reflecting their activation status.

Cofilin is an actin-related protein associated with actin polymerization which effects cell motility and morphology. Destrin is also involved in actin polymerization. Again, this study shows that cytoskeletal proteins are a major group regulated by UV radiation. As such, they are associated with regenerative mechanisms as well as inflammation observed in vivo [52].

UVA is associated with photoaging and carcinogenesis, the former demonstrated by dermal connective tissue damage. Lamore et al. [53] have used proteomics to identify proteins in cultured human skin fibroblasts treated with UVA. They used both chronic and acute protocols and protein changes were characterised by 2D-DIGE. Of the proteins identified, nucleophosmin and cathepsin B were highlighted. Nucleophosmin was upregulated after UVA treatment, whereas cathepsin B was downregulated. The downregulation of cathepsin B as a representative of lysosomal proteases is associated with the build up of lipofuscin in fibroblasts. This accumulation contributes to photodamage and the ageing process.

The failure to repair thymine dimers is found in xeroderma pigmentosis due to deficiencies in translesion synthesis (TLS) involving specialised DNA polymerase. Yuasa et al. have analysed this further using HeLa cells over expressing DNA polη. After exposure to UV radiation they found polη interacts with Rad18/Rad6 and/or Rev1 protein in chromatin fractions. Rev1 was less abundant than Rad18. They used mass spectrometry and Western blot analysis to define these changes. They concluded that with UV induced DNA damage and the presence of thymine dimers effective DNA repair requires polη, Rad18/Rad6 and Rev1. The interaction of these proteins is initiated through replication fork arrest at the sites of UV induced UV induced DNA damage [54].

UV radiation is well known to induce cutaneous skin tumours. However, to date there are no reports using proteomics to analysis these tumours and their development. Shi et al. did use proteomics to examine the differences between progressive and regressive UV-induced fibrosarcomas [55]. They used cell lines to undertake these investigations. Twenty-five differentially expressed proteins were revealed, including 14-3-3 proteins, heat shock proteins, profilin-1 and a fragment of complement C3. Three secreted proteins were myeloperoxidase, alpha-2-macroglobulin and a vitamin D binding protein.

With progressive tumours only seven proteins stood out, particularly 60S ribosomal proteins, complement C3, galactokinase 1, and transketolase. With regressive tumours there were 18 proteins. The abundant ones were three different heat shock proteins, 14-3-3 proteins (beta/alpha and epsilon), alpha actin, proteasome sub units (alpha type 3 and beta type 2) and profilin. Of particular interest are the 14-3-3 proteins as 14-3-3σ is involved in controlling epidermal proliferation/differentiation by forcing cells to exit the cell cycle. Its down regulation can lead to the immortalisation of keratinocytes [56].

A challenge is to relate these differential protein expression profiles to the biology of these tumours. Heat shock protein is a molecule linked to immune events and hence raises the place of immunity in these regressive tumours. Profilin is a tumour suppressor [57] and hence its presence would favour regression. The limited number of proteins in the progressive tumours remains a puzzle.

Ripken did characterise immune UV-induced squamous cell carcinomas with either progressive or regressive characteristics [58]. Halliday has used these cells to examine immune events associated with these tumours [59, 60]. Proteomics now offers another approach to study these tumours, particularly the differentiation events and death pathways.

8.3 Discussion

The use of proteomics to study the effects of UV radiation on skin is to date limited. Cell lines and reconstructed skin have been used to characterise a range of changes after UV exposure. These include evidence of oxidative damage, the expression of stress-induced heat shock proteins, cytokeratins and cytoskeletal proteins, including actin-binding proteins like cofilin. These reflect cell injury and the regenerative capacity of injured keratinocytes [42, 52].

An important finding of Bertrand-Vallery and co-workers was the expression of enhanced levels of TRIM29 allowing keratinocytes to enter a survival pathway rather than cell death. No doubt these cells would retain UV-induced mutations that trigger the malignant pathway [44].

While upregulation of proteins is prominent in these studies, down regulation of proteins is also a feature as seen with cathepsin B in UVA treated human skin fibroblast cultures [53].

The findings of Yuasa et al. are important showing that UV induced DNA damage for effective repair requires polη, Rad18/Rad6 and Rev1 [54].

In the whole animal these changes in cultured cells are reflected in the regeneration of keratinocytes across 24–72 h and subsequent remodelling of the epidermis [4]. It needs to be stressed that while these proteomics studies of cultured cells provides important information it must be seen in the context of the total skin response to UV radiation which extends over days to weeks and involves not only keratinocytes and fibroblasts, but inflammatory and immune cells, including mast cells. To date we have found no proteomic studies on the proteins produced by these latter cells after UV radiation. Molecules we know are involved in these reactions are histamine, prostaglandins, leukotrienes, cytokines and cis-urocanic acid [2]. It is likely that proteomic studies on the inflammatory and immune cells associated with the events after UV radiation would provide critical information into the molecular events involved and provide a direction for further study.

In the future proteomics could be used to clarify some of the questions involved in photobiology. These include short term versus long term effects of UVB and UVA, skin ageing and the complexity of the malignant process in non-melanoma cancer e.g. squamous and basal cell carcinomas as well as melanoma. The study of Lamore et al. using proteomics has implicated cathepsin B leading to a build up of lipofuscin in fibroblasts in photodamage and the ageing process [53]. This provides a starting point for further studies.

The study of Shi et al. shows the complexity of the range of proteins produced by UV induced progressive and regressive fibrosarcomas [55]. But it does provide a model to study progressive and regressive UV induced squamous cell carcinomas. A real challenge is to analyse human skin tumours and their precursor lesions. The relationship between solar keratosis and squamous cell carcinoma could well be dissected using proteomics. Key questions remain in our understanding of melanoma including its genesis and tumours arising in skin and non-sun exposed skins. Proteomics as well as the many different dermal connective tissue tumours such as dermatofibromas and dermatofibrosarcoma protuberans could explore these questions.

This overview of the effects of UV radiation on skin has revealed limited use of proteomics as an investigative tool in the skin response to UV radiation. We conclude this could provide major insights not only in the acute response to UV-induced skin injury but particularly to ageing and the challenges of a diverse range of skin tumours, epithelial, melanocyte and those of the dermal connective tissue.

References

1. Woods GM, Malley RC, Muller HK (2005) The skin immune system and the challenge of tumour immunosurveillance. Eur J Dermatol 15:63–69
2. Clydesdale GJ, Dandie GW, Muller HK (2001) Ultraviolet light induced injury: immunological and inflammatory effects. Immunol Cell Biol 79:547–568
3. Beissert S, Schwarz A, Schwarz T (2006) Regulatory T cells. J Invest Dermatol 126:15–24
4. Ouhtit A, Muller HK, Davis DW et al (2000) Temporal events in skin injury and the early adaptive responses in ultraviolet-irradiated mouse skin. Am J Pathol 156:201–207
5. Gregory CD, Devitt A (2004) The macrophage and the apoptotic cell: an innate immune interaction viewed simplistically? Immunology 113:1–14
6. Xu W, Roos A, Schlagwein N et al (2006) IL-10-producing macrophages preferentially clear early apoptotic cells. Blood 107:4930–4937
7. Dominiak M, Kries RV, Ritt E et al (1989) Ultrastructural morphometric analysis of epidermal melanin distribution following irradiation with UVA or UVB. J Invest Dermatol 92:421
8. Hawk JLM (1982) The effects of sunlight on skin. Practitioner 226:1258
9. Marks R, Rennie G, Selwood TS (1988) Malignant transformation of solar keratoses to squamous cell carcinoma. Lancet 1:795–797
10. Payling Wright G (1958) An introduction to pathology. Longmans, London
11. Corona R, Dogliotti E, D'Errico M et al (2001) Risk factors for basal cell carcinoma in a Mediterranean population: role of recreational sun exposure early in life. Arch Dermatol 137:1162–1168
12. Tilli CM, Van Steensel MA, Krekels GA et al (2005) Molecular aetiology and pathogenesis of basal cell carcinoma. Br J Dermatol 152:1108–1124
13. Green A, Whiteman D, Frost C et al (1999) Sun exposure, skin cancers and related skin conditions. J Epidemiol 9:S7–S13
14. Whiteman DC, Watt P, Purdie DM et al (2003) Melanocytic nevi, solar keratoses, and divergent pathways to cutaneous melanoma. J Natl Cancer Inst 95:806–812
15. Whiteman DC, Pavan WJ, Bastian BC (2011) The melanomas: a synthesis of epidemiological, clinical, histopathological, genetic, and biological aspects, supporting distinct subtypes, causal pathways, and cells of origin. Pigment Cell Melanoma Res 24(5):879–897
16. Noonan FP, Recio JA, Takayama H et al (2001) Neonatal sunburn and melanoma in mice. Nature 413:271–272
17. Noonan FP, Dudek J, Merlino G et al (2003) Animal models of melanoma: an HGF/SF transgenic mouse model may facilitate experimental access to UV initiating events. Pigment Cell Res 16:16–25
18. Hacker E, Irwin N, Muller HK et al (2005) Neonatal ultraviolet radiation exposure is critical for malignant melanoma induction in pigmented Tpras transgenic mice. J Invest Dermatol 125:1074–1077
19. Hacker E, Muller HK, Irwin N et al (2006) Spontaneous and UV radiation-induced multiple metastatic

20. Dewar AL, Doherty KV, Woods GM et al (2001) Acquisition of immune function during the development of the Langerhans cell network in neonatal mice. Immunology 103:61–69
21. Muller HK, Malley RC, McGee HM et al (2008) Effect of UV radiation on the neonatal skin immune system- implications for melanoma. Photochem Photobiol 84:47–54
22. Halliday GM, Muller HK (1987) Sensitization through carcinogen induced Langerhans cell deficient skin activates long lived suppressor cells for both cellular and humoral immunity. Cell Immunol 109: 206–221
23. Maeda A, Beissert S, Schwarz T et al (2008) Phenotypic and functional characterization of ultraviolet radiation-induced regulatory T cells. J Immunol 180:3065–3071
24. Walterscheid JP, Nghiem DX, Kazimi N et al (2006) Cis-urocanic acid, a sunlight-induced immunosuppressive factor, activates immune suppression via the 5-HT2A receptor. Proc Natl Acad Sci USA 103:17420–17425
25. Ansel JC, Kaynard AH, Armstrong CA et al (1996) Skin-nervous system interactions. J Invest Dermatol 106:198–204
26. Foreman JC (1987) Substance P and calcitonin gene-related peptide: effects on mast cells and in human skin. Int Arch Allergy Appl Immunol 82:366–371
27. van der Heijden MW, van der Kleij HPM, Rocken M et al (2005) Mast cells. In: Bos J (ed) Skin immune system, 3rd edn. CRC Press, Boca Raton, pp 237–261
28. Grimbaldeston MA, Nakae S, Kalesnikoff J et al (2007) Mast cell-derived interleukin 10 limits skin pathology in contact dermatitis and chronic irradiation with ultraviolet B. Nat Immunol 8:1095–1104
29. Chacon-Salinas R, Limon-Flores AY, Chavez-Blanco AD et al (2011) Mast cell-derived IL-10 suppresses germinal center formation by affecting T follicular helper cell function. J Immunol 186:25–31
30. Loser K, Mehling A, Loeser S et al (2006) Epidermal RANKL controls regulatory T-cell numbers via activation of dendritic cells. Nat Med 12:1372–1379
31. Byrne SN, Halliday GM (2005) B cells activated in lymph nodes in response to ultraviolet irradiation or by interleukin-10 inhibit dendritic cell induction of immunity. J Invest Dermatol 124:570–578
32. Byrne SN, Ahmed J, Halliday GM (2005) Ultraviolet B but not A radiation activates suppressor B cells in draining lymph nodes. Photochem Photobiol 81:1366–1370
33. Rana S, Byrne SN, MacDonald LJ et al (2008) Ultraviolet B suppresses immunity by inhibiting effector and memory T cells. Am J Pathol 172:993–1004
34. Kripke ML, Cox PA, Bucana C et al (1996) Role of DNA damage in local suppression of contact hypersensitivity in mice by UV radiation. Exp Dermatol 5:173–180
35. Nishigori C, Yarosh DB, Ullrich SE et al (1996) Evidence that DNA damage triggers interleukin 10 cytokine production in UV-irradiated murine keratinocytes. Proc Natl Acad Sci USA 93:10354–10359
35a. Fisher MS, Kripke ML (1977) Systemic alteration induced in mice by ultraviolet light irradiation and its relationship to ultraviolet carcinogenesis. Proc Natl Acad Sci U S A 74:1688–1692
36. Muller HK, Halliday GM, Woods GM (2004) The skin immune system and tumor immunosurveillance. In: Bos JD (ed) Skin immune system, 3rd edn. CRC Press, Boca Raton, pp 475–496
37. Huang CM, Xu H, Wang CC et al (2005) Proteomic characterization of skin and epidermis in response to environmental agents. Expert Rev Proteomics 2: 809–820
38. Scott DK, Lord R, Muller HK et al (2007) Proteomics identifies enhanced expression of stefin A in neonatal murine skin compared with adults: functional implications. Br J Dermatol 156:1156–1162
39. Ouhtit A, Muller HK, Gorny A et al (2000) UVB-induced experimental carcinogenesis: dysregulation of apoptosis and p53 signalling pathway. Redox Rep 5:128–129
40. Rittie L, Fisher GJ (2002) UV-light-induced signal cascades and skin aging. Ageing Res Rev 1: 705–720
41. Svobodova A, Walterova D, Vostalova J (2006) Ultraviolet light induced alteration to the skin. Biomed Pap Med Fac Univ Palacky Olomouc Czech Repub 150:25–38
42. Bertrand-Vallery V, Boilan E, Ninane N et al (2010) Repeated exposures to UVB induce differentiation rather than senescence of human keratinocytes lacking p16(INK-4A). Biogerontology 11: 167–181
43. Dickson MA, Hahn WC, Ino Y et al (2000) Human keratinocytes that express hTERT and also bypass a p16(INK4a)-enforced mechanism that limits life span become immortal yet retain normal growth and differentiation characteristics. Mol Cell Biol 20: 1436–1447
44. Bertrand-Vallery V, Belot N, Dieu M et al (2010) Proteomic profiling of human keratinocytes undergoing UVB-induced alternative differentiation reveals TRIpartite Motif Protein 29 as a survival factor. PLoS One 5:e10462
45. Del Bino S, Vioux C, Rossio-Pasquier P et al (2004) Ultraviolet B induces hyperproliferation and modification of epidermal differentiation in normal human skin grafted on to nude mice. Br J Dermatol 150: 658–667
46. Sano T, Kume T, Fujimura T et al (2009) Long-term alteration in the expression of keratins 6 and 16 in the epidermis of mice after chronic UVB exposure. Arch Dermatol Res 301:227–237
47. Li D, Turi TG, Schuck A et al (2001) Rays and arrays: the transcriptional program in the response of human epidermal keratinocytes to UVB illumination. FASEB J 15:2533–2535

48. Toivola DM, Zhou Q, English LS et al (2002) Type II keratins are phosphorylated on a unique motif during stress and mitosis in tissues and cultured cells. Mol Biol Cell 13:1857–1870
49. Martindale JL, Holbrook NJ (2002) Cellular response to oxidative stress: signaling for suicide and survival. J Cell Physiol 192:1–15
50. Jinlian L, Yingbin Z, Chunbo W (2007) p38^{MAPK} in regulating cellular responses to ultraviolet radiation. J Biomed Sci 14:303–312
51. Sheehan JM, Young AR (2002) The sunburn cell revisited: an update on mechanistic aspects. Photochem Photobiol Sci 1:365–377
52. Hensbergen P, Alewijnse A, Kempenaar J et al (2005) Proteomic profiling identifies an UV-induced activation of cofilin-1 and destrin in human epidermis. J Invest Dermatol 124:818–824
53. Lamore SD, Qiao S, Horn D et al (2010) Proteomic identification of cathepsin B and nucleophosmin as novel UVA-targets in human skin fibroblasts. Photochem Photobiol 86:1307–1317
54. Yuasa MS, Masutani C, Hirano A et al (2006) A human DNA polymerase eta complex containing Rad18, Rad6 and Rev1; proteomic analysis and targeting of the complex to the chromatin-bound fraction of cells undergoing replication fork arrest. Genes Cells 11:731–744
55. Shi Y, Elmets CA, Smith JW et al (2007) Quantitative proteomes and in vivo secretomes of progressive and regressive UV-induced fibrosarcoma tumor cells: mimicking tumor microenvironment using a dermis-based cell-trapped system linked to tissue chamber. Proteomics 7:4589–4600
56. Cianfarani F, Bernardini S, De Luca N et al (2011) Impaired keratinocyte proliferative and clonogenic potential in transgenic mice overexpressing 14-3-3sigma in the epidermis. J Invest Dermatol 131:1821–1829
57. Roy P, Jacobson K (2004) Overexpression of profilin reduces the migration of invasive breast cancer cells. Cell Motil Cytoskeleton 57:84–95
58. Hostetler LW, Romerdahl CA, Kripke ML (1989) Specificity of antigens on UV radiation-induced antigenic tumor cell variants measured in vitro and in vivo. Cancer Res 49:1207–1213
59. Halliday GM, Le S (2001) Transforming growth factor-beta produced by progressor tumors inhibits, while IL-10 produced by regressor tumors enhances, Langerhans cell migration from skin. Int Immunol 13:1147–1154
60. Lucas AD, Halliday GM (1999) Progressor but not regressor skin tumours inhibit Langerhans' cell migration from epidermis to local lymph nodes. Immunology 97:130–137

Effects of Ultraviolet Radiation on Skin Cell Proteome

Riikka Pastila

Abstract

Ultraviolet (UV) radiation is known to cause both positive and negative health effects for humans. The synthesis of vitamin D is one of the rare beneficial effects of UV. The negative effects, such as sunburn and premature photoaging of the skin, increase the risk of skin cancer, which is the most detrimental health consequence of UV radiation. Although proteomics has been extensively applied in various areas of the biomedical field, this technique has not been commonly used in the cutaneous biology. Proteome maps of human keratinocytes and of murine skin have been established to characterize the cutaneous responses and the age-related differences. There are very few publications, in which proteomic techniques have been utilized in photobiology and hence there is no systematic research data available of the UV effects on the skin proteome. The proteomic studies have mainly focused on the UV-induced photoaging, which is the consequence of the long-term chronic UV exposure. Since the use of proteomics has been very narrow in the photobiology, there is room for new studies. Proteomics would offer a cost-effective way to large-scale screen the possible target molecules involved in the UV-derived photodamage, especially what the large-scale effects are after the acute and chronic exposure on the different skin cell populations.

Keywords

Proteome • Skin • Non-ionizing radiation • Ultraviolet radiation • Photo aging • Skin cancer • Melanoma

R. Pastila (✉)
STUK – Radiation and Nuclear Safety Authority,
P.O. Box 14, Helsinki FI-00881, Finland
e-mail: Riikka.Pastila@stuk.fi

9.1 The Effects of UV on the Skin

Sunlight is the most prominent source of ultraviolet (UV) radiation. Ultraviolet radiation spans a wavelength of 100–400 nanometers (nm), being both non-ionizing and non-visible. The terrestrial spectrum of solar UV radiation consists, depending on latitude and season of the year, of 1–5% ultraviolet B radiation (280–320 nm), whereas the majority of the radiation reaching the Earth's surface belongs in the ultraviolet A (320–400 nm) region. Several artificial UV sources, mainly the UV lamps, have been developed for the various purposes, such as for the cosmetic tanning of the skin, for the therapeutic use in the phototherapy, and for the sanitation and germicidal use.

The acute signs of UV exposure are pigmentation (tanning), erythema (sunburn) and the synthesis of vitamin D, which is one of the rare beneficial health effects of UV radiation. Chronic exposure to UV radiation causes premature skin aging and increases the risk of skin cancer. The carcinogenic potential of UV is associated with its ability to suppress the cell-mediated immune responses. Primarily this phenomenon was supposed to prevent the development of excessive inflammation. However, UV-induced immunosuppression may comprise a major risk factor for the development of skin cancer by allowing cancer cells to escape from the immunosurveillance [1].

Skin consists of two major layers. The outermost layer, epidermis, is made up of the stratified squamous epithelium providing the protective barrier against environmental stress. It is formed mainly of keratinocytes, Langerhans cells, and melanocytes. Beneath the epidermis lies a dermis layer, composed of fibroblasts and the connective tissue. Dermis offers skin the strength and elasticity and it also contains the blood capillary system that provides nutrients, and aids in the body's temperature regulation and in metabolic waste product removal.

UVA and UVB radiation have different biological effects on skin. UVB wavelengths are estimated to contribute 80% of the harmful effects of exposure to the sun [2]. Photon energy grows along with the shorter wavelength and thus, UVB wavelengths are more potent in initiating skin carcinogenesis through DNA damage than UVA. UVB mutagenesis is characterized by a high frequency of CC to TT transitions in a DNA strand. These CC to TT tandem mutations are considered as the UVB signature mutations, i.e. so called pyrimidine hot-spots [3, 4]. The genotoxicity of UVA occurs mainly through an indirect photosensitization process via the generation of reactive oxygen species (ROS) that are capable of inducing oxidative DNA damage and mutations [5, 6].

UV radiation is considered as a complete carcinogen of non-melanoma skin cancers (NMSC) by initiating and promoting the carcinogenesis of squamous cell carcinoma (SCC) and basal cell carcinoma (BCC). A direct correlation between UVB-induced pyrimidine hot spot mutations in the tumor suppressor protein p53 and the onset of SCC and BCC provides direct evidence for the mutagenic role of UVB in skin carcinogenesis [7–9]. In the melanoma development the main etiological risk factor is the UV radiation, although hereditary reasons also play a notable role in the progression of malignant melanoma. As with the NMSC, individuals sensitive to the sun, who do not tan and burn easily, are at the greatest risk.

Photoaging, i.e. premature aging of the photodamaged skin, is a result of chronic exposure to UV radiation. The clinical signs of photoaging are dryness, roughness, deep wrinkles, irregular pigmentation, elastosis and telangiectasia, which appear in areas heavily exposed to the sun, such as the face, neck and the upper extremities. Photoaging predisposes to the formation for solar keratosis, which is considered a precursor of squamous cell carcinoma. Although UVB radiation is mainly responsible for DNA damage and eventually skin cancer formation, UVA is considered a major factor in the process of skin photoaging. UVA-derived reactive oxygen species lead to the accumulation of disorganized elastic fibers in the dermal compartment causing also loss of interstitial collagen, the main component of the dermal connective tissue [10]. Indeed, the histological hallmark of photoaging, called as solar elastosis, is the massive accumulation of atypical elastotic material in the upper and middle dermis.

9.2 Proteome Analysis of the Skin

9.2.1 Mapping the Human Skin Cells

Proteomics provides nowadays an effective tool to analyze simultaneously the expression profiles of the several proteins. The high throughput protocols, e.g. two-dimensional polyacrylamide gel electrophoresis (2D-PAGE), fluorescent two dimensional difference in-gel electrophoresis (2D-DIGE), matrix-assisted laser desorption/ionization time-of-flight mass spectrometry (MALDI-TOF MS), and various chromatography techniques, like liquid chromatography (LC), or tandem mass spectrometry (MS/MS) allow the large-scale analysis of the whole proteome and enable the profiling of different protein isoforms and post-translational modifications, such as phosphorylation and glycosylation.

Although proteomics has been extensively applied in various areas of the biomedical field, this technique has not been commonly used in the cutaneous biology [11, 12]. In 1990s, Celis et al. gathered novel information from the epidermal keratinocytes of the biopsied human skin using 2D-PAGE, microsequencing, and mass spectrometry [13–16]. They established a keratinocyte 2D-PAGE database, in which over 1,000 keratinocyte-derived proteins were identified. This database has been used in profiling bladder squamous cell carcinoma, which resembles closely skin keratinocytes both in morphology and protein expression patterns [17]. Quantitative analysis of protein expression profiles of human epidermis, biopsied from the elderly people, was also generated by Celis' group [18]. Skin proteome was identified by matching the gels with the master 2D gel image of human keratinocyte reference map described above, and selected proteins were confirmed by immunoblotting and/or mass spectrometry. Quantitative analysis of 172 proteins showed that the majority of them (148) remained unaffected by the aging process, but the 22 deregulated proteins, like manganese-superoxide dismutase (Mn-SOD) and heat shock protein 60 (HSP60), support the notion that aging is linked with increased oxidative stress, and the concomitant apoptosis. This is in accordance with the notion that aging is related with the increased oxidative stress, exacerbated by increased production of reactive oxygen species (ROS).

Other proteome maps of skin structures have also been generated. For example a proteome map of a human fibroblast cell line has been established to determine collagen and collagen-related proteins, and to screen the metabolic disorders of the skin [19]. In addition, human epidermal plasma membrane has been characterized by 2D-LC and MS/MS [20]. In that study 57.3% of the identified proteins were assigned as integral membrane or membrane-associated proteins, such as intercellular adhesion proteins and gap junction proteins.

There are also few publications about proteome profiling of special skin organelles. Mitochondrial proteins are essential in metabolics and in regulating apoptosis, and number of disorders is known to be related to mitochondrial dysfunction. Mitochondria from human fibroblasts were characterized by nano-LC-MS/MS analysis of iTRAQ-labeled peptides to study the relative amounts of mitochondrial protein levels, and to identify the metabolic imbalance and cellular stress [21]. Melanosomes [22] and lamellar bodies of the epidermis [23] were identified by 2D-MS/MS, or by nano-LC-MS/MS, respectively, in a scope to study further the biogenesis of lysosome-related organelles. These organelles are responsible for many critical functions in the skin cells and proteome analysis will help and improve the understanding of their biological function.

9.2.2 Mapping the Murine Skin

To characterize the epidermal and dermal responses for the environmental changes, a reference proteome map from BALB/c abdominal area was established by 2-DE and mass spectrometry [24]. Out of identified 34 proteins, 25 proteins were found to be expressed in the epidermal compartment, whereas 9 proteins were expressed more predominantly in the dermis. Proteins were involved for example in the stress response, apoptosis, growth inhibition, energy metabolism, cholesterol transport and scavenging the free radicals. In the future, this map might help to identify the protein targets for the prediction and prevention of environmentally induced skin conditions, such as skin allergy, carcinogenesis and microbial infections [12, 25].

Scott et al. studied age-related differences in the mouse model. They identified 179 differentially expressed proteins in neonatal BALB/c mice skin as compared to the adult tissue [26]. One such protein was Stefin A, which was abundant in the neonatal skin, but its expression decreased with age, suggesting a functional change during development. Since Stefin A is normally up-regulated in the proliferative diseases, like psoriasis, authors postulated that it is likely that this protein could provide a useful target for diseases of abnormal proliferative conditions, including cancer [26, 27].

9.3 Proteome Analysis of Skin Cells After UV Exposure

Proteomics is not widely used method in the photobiology. There are very few publications, in which proteomic techniques have been utilized in analyzing the UV-derived effects, and hence there is no systematic research data available of the UV effects on the skin proteome.

There is only one study so far, where the *acute* photodamage has been studied among the benign skin cells. Photodamage was induced by a single UVB exposure (100 mJ/cm^2) in the dermal fibroblasts by 2D-PAGE/MS, and obtained protein spots were confirmed by qPCR and Western blot analysis [28]. In the UVB-treated cells, 18 differentially expressed proteins were identified, from which receptor-interacting protein (RIP) and vimentin were significantly up-regulated. RIP plays a role in triggering apoptosis, whereas vimentin is a vital cytoskeletal protein in fibroblasts, protecting them from physical injury and DNA damage. Overexpression of vimentin can also delay a cell death. Authors speculated that RIP may represent a key regulatory step in triggering cell death after UVB exposure, whereas vimentin may contribute to the resistance of cells to UVB-induced damage [28].

The proteomic studies have mainly focused on the UV-induced photoaging, which is the consequence of the long-term *chronic* UV exposure. Bertrand-Vallery et al. studied the effect of the multiple UVB radiation doses (8×300 mJ/cm^2) on the keratinocytes by 2D-DIGE [29]. Sixty-nine differentially expressed proteins, which were involved in the keratinocyte differentiation and survival, were identified by LC-MS/MS. Out of the identified proteins the most interesting candidate for further analysis was TRIparite Motif Protein 29 (TRIM29), validated by Western blot, immunochemistry, and RT-PCR. It was found to be expressed very abundant in keratinocytes and reconstructed epidermis. Originally TRIM29 was discovered from the cells of Ataxia Telangienctasia patients, and it is known to partially suppress the sensitivity to ionizing radiation [30]. Bertrand-Vallery et al. suggested that TRIM29 participates in the survival of differentiating keratinocytes and may allow them to enter a protective alternative differentiation process, rather than die massively after UV-induced stress. This finding is a novel mechanism that allows the survival of keratinocytes as they migrate and differentiate in the skin [29].

When the chronic UVA exposure-induced alterations were studied by 2D-DIGE/MS in the dermal fibroblast cell culture, the most pronounced protein undergoing UVA-induced down-regulation was cathepsin B, verified by Western blot analysis [31]. The used UVA regimen was given as a chronic exposure for 3 weeks and the total UVA dose was 59.4 J/cm^2. Cathepsin B is a lysosomal cysteine-protease, which is known to be inactivated by oxidative-stress related conditions in the cells. Indeed, Lamore et al. showed, that UVA-induced loss of cathepsin B enzymatic activity in fibroblasts was suppressed by antioxidant intervention, and that UVA-irradiation of fibroblasts led to the massive lipofuscin accumulation in lysosomes, most likely due to the impaired cathepsin B activity. Authors discussed the possibility that cathepsin B may be a crucial target of UVA-induced photo-oxidative stress causatively involved in dermal photodamage through the impairment of lysosomal removal of lipofuscin [31].

In addition to the cell culture studies, the albino hairless HR-1 mouse model has also been utilized when studying the photoaging [32]. Mice were irradiated 5 days per week for 7 weeks with the total UVB dose of 2.31 J/cm^2. Proteomic analysis revealed 17 differently expressed proteins in the photoaged skin, for example superoxide dismutase (SOD) and malonialdehyde (MDA). Additionally in this set-up, the antioxidant properties of Korean deciduous tree extract, *Machilus thunbergii Sieb et Zucc* (*M. thunbergii*), widely used in the traditional medicine, was examined. The dorsal skin of mice was treated topically with *M. thunbergii* for 2 h prior to UV irradiation and proteomes from the skin of each study group were analyzed. *M. thunbergii* treatment altered the expression of several proteins. Interestingly, among 15 proteins upregulated by UV-irradiation, 13 proteins were down-regulated after treatment with *M. thunbergii*. Also the thickness of the dorsal skin was significantly decreased in the group that was UV-exposed and treated with *M. thunbergii* extract. This result possibly indicates that *M. thunbergii* might have had antiphotoaging effects [32].

So far, there is only one proteomic study performed with human excised native skin after the chronic UV exposure. When skin samples were irradiated with solar simulated UV radiation for periods of 3 or 4 h (the total cumulative UV dose of 11.25 J/cm^2 and 15 J/cm^2, respectively) and analyzed by 2D-PAGE/MS, oxidative stress markers were seen expressed differentially in the UV irradiated samples and the non-exposed controls [33]. Apart from the proteins known as the UV-related oxidative stress markers, such as HSP27 and MnSOD, authors identified also two novel proteins that were down-regulated after UV exposure. Further analysis revealed that these proteins were the de-phosphorylated forms of cofilin-1 and destrin, which are known to be actin-cytoskeleton modulators [33].

9.4 Conclusions

As a summary, so far the UV-induced alterations found by the proteomic approach have mainly been involved in the oxidative stress-related changes. Since the use of proteomics has been very narrow in the photobiology, there is room for new studies. Proteomics would also offer a cost-effective way to large-scale screen the possible target molecules involved in the UV-derived photodamage. Both in vitro and especially in vivo experimental set-ups are required to better answer the question what the large-scale effects are after the acute and chronic exposure on the different skin cell populations, and what novel target molecules can be found on the skin proteome after exposing the skin cells with the different wavelengths of UV radiation.

Finally, since the skin cancer formation is the most detrimental health hazard deriving from the UV radiation, proteomics might offer a useful way to study photocarcinogenesis. Malignant melanoma is associated with poor prognosis, especially if the tumor has progressed into the metastatic phase. Thus, early detection would improve substantially patient survival. Proteomics also appears to be an ideal choice for the discovery of new melanoma biomarkers in humans [34]. Proteins specifically secreted by tumor cells, for example proteins shed into the blood, may serve as early cancer biomarkers. Several proteomic approaches have been utilized in the search for clinically relevant biomarkers in melanoma, but so far the results have been relatively limited. Human melanoma proteomic studies have been reviewed in a detailed manner in the excellent review article written by Sabel et al. [35].

References

1. Ullrich SE (2005) Mechanisms underlying UV-induced immune suppression. Mutat Res 571: 185–205

2. Diffey BL (1998) Ultraviolet radiation and human health. Clin Dermatol 16:83–89
3. Matsumura Y, Ananthaswamy HN (2004) Toxic effects of ultraviolet radiation on the skin. Toxicol Appl Pharmacol 195:298–308
4. Melnikova VO, Ananthaswamy HN (2005) Cellular and molecular events leading to the development of skin cancer. Mutat Res 571:91–106
5. Cadet J, Sage E, Douki T (2005) Ultraviolet radiation-mediated damage to cellular DNA. Mutat Res 571:3–17
6. Pfeifer GP, You YH, Besaratinia A (2005) Mutations induced by ultraviolet light. Mutat Res 571:19–31
7. Brash DE, Rudolph JA, Simon JA, Lin A, McKenna GJ, Baden HP, Halperin AJ, Ponten J (1991) A role for sunlight in skin cancer: UV-induced p53 mutations in squamous cell carcinoma. Proc Natl Acad Sci USA 88:10124–10128
8. Ziegler A, Jonason AS, Leffell DJ, Simon JA, Sharma HW, Kimmelman J, Remington L, Jacks T, Brash DE (1994) Sunburn and p53 in the onset of skin cancer. Nature 372:773–776
9. Ziegler A, Jonason A, Simon J, Leffell D, Brash DE (1996) Tumor suppressor gene mutations and photocarcinogenesis. Photochem Photobiol 63:432–435
10. Kondo S (2000) The roles of cytokines in photoaging. J Dermatol Sci 23(Suppl 1):S30–S36
11. Jansen BJ, Schalkwijk J (2003) Transcriptomics and proteomics of human skin. Brief Funct Genomic Proteomic 1:326–341
12. Huang CM, Elmets CA, Van Kampen KR, DeSilva TS, Barnes S, Kim H, Tang DC (2005) Prospective highlights of functional skin proteomics. Mass Spectrom Rev 24:647–660
13. Celis JE, Rasmussen HH, Madsen P, Leffers H, Honore B, Dejgaard K, Gesser B, Olsen E, Gromov P, Hoffmann HJ et al (1992) The human keratinocyte two-dimensional gel protein database (update 1992): towards an integrated approach to the study of cell proliferation, differentiation and skin diseases. Electrophoresis 13:893–959
14. Celis JE, Rasmussen HH, Olsen E, Madsen P, Leffers H, Honore B, Dejgaard K, Gromov P, Hoffmann HJ, Nielsen M et al (1993) The human keratinocyte two-dimensional gel protein database: update 1993. Electrophoresis 14:1091–1198
15. Celis JE, Rasmussen HH, Olsen E, Madsen P, Leffers H, Honore B, Dejgaard K, Gromov P, Vorum H, Vassilev A et al (1994) The human keratinocyte two-dimensional protein database (update 1994): towards an integrated approach to the study of cell proliferation, differentiation and skin diseases. Electrophoresis 15:1349–1458
16. Celis JE, Rasmussen HH, Gromov P, Olsen E, Madsen P, Leffers H, Honore B, Dejgaard K, Vorum H, Kristensen DB et al (1995) The human keratinocyte two-dimensional gel protein database (update 1995): mapping components of signal transduction pathways. Electrophoresis 16:2177–2240
17. Celis JE, Ostergaard M, Jensen NA, Gromova I, Rasmussen HH, Gromov P (1998) Human and mouse proteomic databases: novel resources in the protein universe. FEBS Lett 430:64–72
18. Gromov P, Skovgaard GL, Palsdottir H, Gromova I, Ostergaard M, Celis JE (2003) Protein profiling of the human epidermis from the elderly reveals up-regulation of a signature of interferon-gamma-induced polypeptides that includes manganese-superoxide dismutase and the p85beta subunit of phosphatidylinositol 3-kinase. Mol Cell Proteomics 2:70–84
19. Oh JE, Krapfenbauer K, Lubec G (2004) Proteomic identification of collagens and related proteins in human fibroblasts. Amino Acids 27:305–311
20. Blonder J, Terunuma A, Conrads TP, Chan KC, Yee C, Lucas DA, Schaefer CF, Yu LR, Issaq HJ, Veenstra TD, Vogel JC (2004) A proteomic characterization of the plasma membrane of human epidermis by high-throughput mass spectrometry. J Invest Dermatol 123:691–699
21. Palmfeldt J, Vang S, Stenbroen V, Pedersen CB, Christensen JH, Bross P, Gregersen N (2009) Mitochondrial proteomics on human fibroblasts for identification of metabolic imbalance and cellular stress. Proteome Sci 7:20
22. Chi A, Valencia JC, Hu ZZ, Watabe H, Yamaguchi H, Mangini NJ, Huang H, Canfield VA, Cheng KC, Yang F, Abe R, Yamagishi S, Shabanowitz J, Hearing VJ, Wu C, Appella E, Hunt DF (2006) Proteomic and bioinformatic characterization of the biogenesis and function of melanosomes. J Proteome Res 5:3135–3144
23. Raymond AA, de Gonzalez PA, Stella A, Ishida-Yamamoto A, Bouyssie D, Serre G, Monsarrat B, Simon M (2008) Lamellar bodies of human epidermis: proteomics characterization by high throughput mass spectrometry and possible involvement of CLIP-170 in their trafficking/secretion. Mol Cell Proteomics 7:2151–2175
24. Huang CM, Foster KW, DeSilva T, Zhang J, Shi Z, Yusuf N, Van Kampen KR, Elmets CA, Tang DC (2003) Comparative proteomic profiling of murine skin. J Invest Dermatol 121:51–64
25. Huang CM, Xu H, Wang CC, Elmets CA (2005) Proteomic characterization of skin and epidermis in response to environmental agents. Expert Rev Proteomics 2:809–820
26. Scott DK, Lord R, Muller HK, Malley RC, Woods GM (2007) Proteomics identifies enhanced expression of stefin A in neonatal murine skin compared with adults: functional implications. Br J Dermatol 156:1156–1162
27. Muller HK, Malley RC, McGee HM, Scott DK, Wozniak T, Woods GM (2008) Effect of UV radiation on the neonatal skin immune system- implications for melanoma. Photochem Photobiol 84:47–54
28. Yan Y, Xu H, Peng S, Zhao W, Wang B (2010) Proteome analysis of ultraviolet-B-induced protein expression in vitro human dermal fibroblasts. Photodermatol Photoimmunol Photomed 26:318–326

29. Bertrand-Vallery V, Belot N, Dieu M, Delaive E, Ninane N, Demazy C, Raes M, Salmon M, Poumay Y, Debacq-Chainiaux F, Toussaint O (2010) Proteomic profiling of human keratinocytes undergoing UVB-induced alternative differentiation reveals TRIpartite Motif Protein 29 as a survival factor. PLoS One 5:e10462
30. Kapp LN, Painter RB (1989) Stable radioresistance in ataxia-telangiectasia cells containing DNA from normal human cells. Int J Radiat Biol 56:667–675
31. Lamore SD, Qiao S, Horn D, Wondrak GT (2010) Proteomic identification of cathepsin B and nucleophosmin as novel UVA-targets in human skin fibroblasts. Photochem Photobiol 86:1307–1317
32. Uhm YK, Jung KH, Bu HJ, Jung MY, Lee MH, Lee S, Lee S, Kim HK, Yim SV (2010) Effects of *Machilus thunbergii Sieb et Zucc* on UV-induced photoaging in hairless mice. Phytother Res 24: 1339–1346
33. Hensbergen P, Alewijnse A, Kempenaar J, van der Schors RC, Balog CA, Deelder A, Beumer G, Ponec M, Tensen CP (2005) Proteomic profiling identifies an UV-induced activation of cofilin-1 and destrin in human epidermis. J Invest Dermatol 124: 818–824
34. Rezaul K, Wilson LL, Han DK (2008) Direct tissue proteomics in human diseases: potential applications to melanoma research. Expert Rev Proteomics 5: 405–412
35. Sabel MS, Liu Y, Lubman DM (2011) Proteomics in melanoma biomarker discovery: great potential, many obstacles. Int J Proteomics 2011:181890

Index

A
Acute effects, 63, 64, 112
Apoptosis, 38, 39, 51, 56, 64, 80, 103, 112, 114, 115, 123, 124

B
Biomarker, 3, 11, 12, 24, 25, 43, 45, 63, 64, 66–68, 70–73, 76–81, 89, 91, 92, 94–96, 109, 125
Bystander effect, 56

C
Cardiovascular system, 41
Cytokine, 41, 42, 45, 50–52, 54, 56, 63–65, 68, 72, 77, 78, 80, 91, 94, 112–114, 116

D
2DE, 92, 94, 104, 108, 124
2DE-DIGE, 102, 104
Delayed effects, 52, 91
2DGE. *See* Two-dimensional gel electrophoresis (2DGE)
Diagnostic, 66, 71, 72, 76, 79, 113
Dosimetry, 63, 76, 89, 91, 93, 102

E
EA.hy926, 37, 38
Electromagnetic fields (EMFs), 101–105, 107, 108, 112
Electrospray ionization (ESI), 2, 5–7
ELF-MFs. *See* Extremely low frequency magnetic fields (ELF-MFs)
EMF. *See* Electromagnetic fields (EMFs)
Endothelial cells, 37, 38, 42, 50, 52, 54, 55, 65, 104
ESI. *See* Electrospray ionization (ESI)
Extracellular matrix (ECM), 50–57
Extremely low frequency magnetic fields (ELF-MFs), 107–109

F
Formalin-fixed, 43–45

G
Glycosylation, 3, 8, 66, 68, 79, 123

H
Heart, 18, 41–44, 65, 71
Heat-shock proteins, 108, 115, 116, 123
Hypoxia, 51, 52, 54, 65

I
Immune system, 40, 50, 81, 112, 113
Immuno suppression, 113, 114, 122
Inflammation, 41, 52, 54, 63, 75, 78–81, 96, 112, 114, 115, 122
Ionizing radiation, 24, 37–45, 55, 57, 62–65, 76, 77, 80–82, 88–91, 94–96, 124
Isotope labeling, 3, 21, 24, 25

K
Keratinocyte, 112–116, 122–124
Kidney, 53, 56, 65, 90, 91, 93–95, 103

L
Langerhans cells (LC), 113, 122
Lymphocytes, 40, 50, 64, 79, 91, 97
Lysis, 52, 54–55, 57

M
MALDI. *See* Matrix-assisted laser desorption/ionization (MALDI)
Mascot, 3, 14, 15, 18, 19
Mass spectrometry (MS), 1–25, 38, 43, 67–70, 73–80, 91–96, 108, 109, 116, 123–125
Matrix-assisted laser desorption/ionization (MALDI), 2, 38, 73, 75–77, 92, 123
Melanoma, 113, 117, 122, 125
Metastasis, 113
Mitochondria, 42–43, 123

N

Nanoscale reversed-phase liquid chromatography (nanoLC), 3, 5, 7, 24, 75, 78, 123
Non-ionizing radiation, 121
Normal tissue toxicity, 72
Nucleolar proteome, 40–41

O

Omics, 65, 81

P

Paraffin-embedded, 24, 43–45
Peptide identification, 3, 4, 8, 12–21, 25
Peptide sequencing, 12, 16
Phosphorylation, 3, 8, 11, 15, 40, 42, 56, 66, 102, 104, 115, 123
Photoaging, 115, 122, 124, 125
Plasma, 45, 50, 61–82, 90, 91, 94–96, 123
Plasma proteins, 66, 72, 73, 81
Prognostic, 66, 78, 79
Protein activity, 102
Proteinase inhibitors, 96
Proteinases, 96
Protein expression, 43, 45, 64, 66, 81, 102–105, 107–109, 116, 123
Proteolysis, 3, 50, 51, 53, 54, 57, 92
Proteome, 2–4, 6, 9, 11, 14, 18, 20, 21, 23, 25, 37–45, 49–57, 64–71, 76, 79, 87–97, 101–105, 108, 109, 111–117, 121–125

R

Radiation injury, 53, 55, 63, 64, 71, 72, 80, 81, 88–91, 95, 96
Radiofrequency-modulated, 101–105
Radiological terrorism, 97
Radiotherapy, 41, 52, 55–57, 63–65, 72, 76–80
Renal, 54, 89, 90, 94–96
RF-EMF, 101–105

S

Sequence database, 3, 4, 9, 15, 21, 23–25
SEQUEST, 3, 14, 15, 18, 92
Serum, 11, 24, 45, 61–82, 90, 102
Signaling pathways, 64
Skin, 40–41, 52, 53, 64, 65, 72, 75, 78–80, 91, 102, 103, 111–117, 121–125
Skin cancer, 111, 112, 122, 125
Stress proteins, 102
Stress response, 38, 39, 102, 115, 124
Stroma, 49–52, 55, 57, 65
Sunburn cells, 112, 115

T

Tanning, 112, 122
Triage, 63, 64, 71–73, 75, 89, 91, 94, 97
Tumour microenvironment, 49–57
Two-dimensional gel electrophoresis (2DGE), 3, 73, 75, 76, 80, 103, 108, 123

U

Ultraviolet radiation (UV), 8, 111–117, 121–125
UVA, 112, 115–117, 122, 124, 125
UVB, 112–115, 117, 122, 124, 125
UVC, 40, 112

Y

Yeast, 6, 10, 108, 109

Printed by Publishers' Graphics LLC
BT20130321.12.05.184